T0269287

SpringerBriefs in Stem Cells

More information about this series at http://www.springer.com/series/10206

Indumathi Somasundaram

Endometrial Stem Cells and Its Potential Applications

Indumathi Somasundaram
Department of Stem cell and Regenerative
Medicine
Centre for Interdisciplinary Research
D.Y. Patil University
Kolhapur, India

ISSN 2192-8118 ISSN 2192-8126 (electronic)
SpringerBriefs in Stem Cells
ISBN 978-81-322-2744-1 ISBN 978-81-322-2746-5 (eBook)
DOI 10.1007/978-81-322-2746-5

Library of Congress Control Number: 2016930395

Springer New Delhi Heidelberg New York Dordrecht London

Printed on acid-free paper

Springer (India) Pvt. Ltd. is part of Springer Science+Business Media (www.springer.com)

I dedicate this book to my sweet little kid.
Akshaya K Rahul

Foreword

This text presents an excellent update of the latest findings in endometrial biology. This book begins with a basic description of the anatomy, physiology, and pathology of the uterus and, in particular, the endometrium. Subsequently, the main focus becomes the role of endometrial stem cells, where both their positive and negative attributes are described. Specifically, the book describes their role in the physiology and development of the endometrium throughout a woman's life, as well as their role in various endometrial disorders. Moreover, the potential use of these cells in clinical settings is addressed.

The field of stem cells has become among the most intensely studied areas in the last decade. Stem cells have the innate property to reproduce and differentiate into other cells. This leads to many possibilities including tissue regeneration and development which can be used for transplant in humans with a range of diseases. Adult stem cells can divide or self-renew indefinitely, allowing them to generate a range of cell types from the originating organ or even regenerate the entire original organ. Additionally, it has now been demonstrated that adult stem cells can be coaxed to differentiate into other tissue cell types.

The latest descriptions on the isolation, phenotypic characterization, and culturing of endometrial stem cells from sources such as menstrual blood, biopsies, or hysterectomies are clearly described. In vitro multi-lineage differentiation strategies demonstrate the high plasticity and potential of endometrial stem cells which can be coaxed into differentiating into ectodermal, mesodermal, and endodermal lineages. The author points out those studies on adult stem cell biology in the uterus lag behind other areas of stem cell research despite the important fact that unlike other organs, the uterus undergoes the most extensive proliferative changes and remodeling in adult mammals. Putative stem or progenitor cells are responsible for the cyclical regeneration of the endometrium functionalis. These stem cells are believed to reside in the endometrium basalis, making this part of the uterus of highest importance. This makes the endometrium a treasure for regenerative medicine as its stem cells are highly regenerative and can likely be used for a wide number of diseases.

While the usefulness of endometrial stem cells in ischemic injury, diabetes, and Parkinson's disease is one of the areas where endometrial progenitor cells have been shown to reverse disease in animal models, these cells can have untoward outcomes due to abnormal proliferation and differentiation.

In summary, the current understanding of stem cells has raised great hopes in the ability to cure disease and enhance the ability to transplant tissues. Ever since the first bone marrow transplant, scientists have been slowly exploring the field which has now exploded with new and exciting findings.

Yale School of Medicine
Yale University
New Haven, CT, USA

Hugh S. Taylor

Preface

Stem cells in recent years have configured regenerative medicine with their wide range of therapeutic potentials. This revolutionary change has increased the demand of stem cells for a myriad of diseases. Their ability to treat so many diseases rests on their unique properties of self-renewal and differentiation. The potentials of stem cells have been demonstrated from various sources such as bone marrow, adipose tissue, cord blood, placenta, amniotic fluid, and so on. Regardless of the ubiquitous presence of stem cells in all tissues, taking those stem cells adaptable for regenerative medicine applications in adequate quantities at the right time is a challenge. This is with regard to the inevitable fact that the frequencies of stem cells, their proliferative capacities and differentiation potentials, as well as phenotypic and immunomodulatory properties have been shown to vary in different sources. Thus, choosing an ideal source of stem cells for treating a multitude of diseases is most imperative. The human endometrium could be considered as an ideal source of stem cells, as it possesses unique attributes that are engaged in the coordinated functions of pregnancy and menstruation throughout reproductive life of women. Besides, endometrial stem cells could also be isolated from the postmenopausal women, suggesting the potential of this stem cell throughout life of women. In a nutshell, human endometrium could be considered as an ideal source of stem cells for treating multitude of diseases in the present and in the near future due to the following reasons: easy to isolate, readily available, non-invasive, longer preservation, highly clonogenic with a higher multi-differentiation ability, and possessing immunomodulatory properties and inherent angiogenic potential. This Springer Brief, thus, indents to address the potential applications of endometrial stem cells as an accessible, noninvasive, rather potent source of stem cells for therapeutic interventions. This is accomplished by demonstrating the unique properties of endometrial stem cells with its in vitro and in vivo applications. The book ends with the emphasis on the threat and challenges of endometrial stem cells in causing gynecological disorders as a base for scientists and researchers to explore the ways of treatment and cures.

Kolhapur, India — Indumathi Somasundaram

Acknowledgement

Many qualified and enthusiastic scientists, researchers, physicians and my students have dedicated time and made precious efforts over the text and concepts presented in this Brief. I owe my very special thanks to Hon'ble Dr. S.H Powar, Vice Chancellor, Dr. V.V. Bhosale, Registrar and Dr. S.P. Kole, Finance officer from D.Y. Patil Univeristy, for their constant support and motivation to complete this book successfully. Dr. Ramesh R Bhonde, from School of Regenerative Medicine, Manipal University, has been a great inspiration in my path. His thought-provoking suggestions with his criticisms and appreciations had always been an encouragement. Dr. Sangeetha Desai and Dr. Vasuda Sawanth, Department of Obstetrics and Gynecology, D.Y. Patil Hospital, Kolhapur has been of constant support in the first part of the book. I owe my gratitude to Dr. B.C. Patil, D.Y Patil hosptal for some of his significant suggestions. I also owe my special thanks to Mr. Pankaj Kaingade, Embryologist for his timely help in creating of illustrations. I also thank my Juniors, Ms. Kanmani Anandan and Mr. Padmanav Behera, for their kind help in illustrations and formatting. Finally, I owe my deep sense of gratitude to the almighty and my family members: my parents, Mr. G. Somasundaram and Mrs. S. Mangalam, My husband, Mr. K.S. Rahul, and my beloved sisters, Mrs. S. Rajalakshmi and Mrs. S. Gayathri, who are the great pillars of my life, invariably showering their blessings, love and prayers forever.

Contents

About the Author

Indumathi Somasundaram is an Assistant Professor of the Dept. of Stem Cells and Regenerative Medicine at the D.Y. Patil University since October 2014. She completed her post graduation in 2008 and worked for 2 years as a Researcher at the Department of Stem Cells, Lifeline RIGID Hospitals Pvt. Ltd. She obtained her PhD in 2014 under the University of Madras with her thesis focusing on reproductive stem cells. She then joined as a Research Associate at the Department of Stem Cell, National Institute of Nutrition (ICMR), Secunderabad, for nearly 1.5 years after submission of her PhD thesis.

During these periods, she gained experience on different kinds of adult stem cells including bone marrow, adipose tissue and endometrium, its in vitro attributes of self renewal and differentiation, and its therapeutic implications/applications. Her areas of interest and expertise include adult stem cells, in specific endometrial stem cell biology, regenerative medicine, reproductive biology, diabetes, cancer stem cells and so on. She has around 20 peer-reviewed publications and four book chapters (Springer Verlag) to her credit and few more in pipeline. She recently edited and published a book on *Stem Cell Therapy for Organ Failure* from Springer Verlag, and had self contributed four chapters in that book. Her main area of research is addressed to the understanding of potentials of human endometrial derived stem cells in regenerative medicine and identifies the possible causes and cures of endometrial disorders and its management. Besides, her present research focus is on molecular profiling of endometrium under normal and disease conditions to find the classical pathway and functions related to this disease to identify a targeted therapeutic treatment approach.

Chapter 1
The Uterus, Endometrium, and Its Derived Stem Cells

Reproduction can be defined as the process by which an organism continues its species. The development of the normal female reproductive tract is a complex process. The paramesonephric ducts arise from the intermediate mesoderm, which are the precursors of the female reproductive organs that includes uterus, fallopian tubes, cervix, and upper vagina. The female reproductive system is designed to carry out several functions. It produces the female egg cells, the oocyte necessary for reproduction. This oocyte gets transported to the site of fertilization. Conception normally occurs in the fallopian tubes. The fertilized egg implants into the uterine wall to begin the initial stages of pregnancy. If fertilization and/or implantation do not take place, the system is designed to menstruate (the monthly shedding of the uterine lining). The female reproductive system produces female sex hormones that maintain the reproductive cycle. The fallopian tubes lead to the uterus, a muscular organ in the pelvic cavity. The inner lining, called the endometrium, thickens with blood and tissue in anticipation of a fertilized egg cell. If fertilization fails to occur, the endometrium degenerates and is shed in the process of menstruation. The endometrium begins to reach full development at puberty and thereafter exhibits remarkable changes during each menstrual cycle. It undergoes further changes before, during, and after pregnancy, during the menopause, and in old age. It undergoes unique tissue remodeling and regeneration, it remains as a dynamic tissue undergoing more than 400 cycles of regeneration, differentiation, and shedding during a woman's lifetime. Thus, endometrium becomes the most important source of study. Based on these dynamic properties, it is ideal that stem cells of the endometrium possess an inherent and higher regenerative potential as compared to other post-natal stem cells.

I. Somasundaram, *Endometrial Stem Cells and Its Potential Applications*,
SpringerBriefs in Stem Cells, DOI 10.1007/978-81-322-2746-5_1

1.1 The Uterus

The uterus, also commonly known as the womb, is a hollow, pear-shaped muscular organ of the female reproductive system, playing an integral role in reproductive life of women. The cavity of the uterus is lined with a special kind of mucous membrane, which is called the endometrium. The uterine cavity communicates through the external os with the vagina and through the fallopian tube with the peritoneal cavity or coelom. The uterus is a remarkably resilient organ where its functions are directed by ovarian-derived steroid hormones. It plays a pivotal role in implantation and in absence of pregnancy, menstruation. It is also an incredibly strong organ that helps in nurturing the fertilized ovum that develops into the fetus until its birth. The walls of the body are much thicker than those of the cervix as they provide for the protection and support of the developing fetus and contain the muscles that propel the fetus out of the mother's body during childbirth. They derive nourishment from blood vessels which develop exclusively for this purpose. Uterus provides structural integrity and support to the bladder, bowel, pelvic bones, and organs as well. It separates the bladder and the bowels. The blood vessels and uterine nerves direct the blood flow to the pelvis and to the external genitalia, including the ovaries, vagina, labia, and clitoris for sexual response.

1.1.1 Development of Uterus

During embryonic development, there is a sexually indifferent stage in which the embryo has the potential to develop either male (testes) or female (ovary) structures (Fig. 1.1). Internally, there are two primitive ducts that can give rise to either the male or the female reproductive tracts. The Wolffian (mesonephric) ducts are more medial. The Müllerian (paramesonephric) ducts are more lateral, but then fuse in the midline more caudally. Sexual differentiation begins with sexual determination, which depends upon the sex chromosomes, X and Y and SRY gene. If the embryo is XY, the *SRY* gene (for sex-determining region of the Y chromosome) will direct the gonads to develop as testes (Fig. 1.2a). If the embryo is XX, in the absence of a Y chromosome and SRY gene, the gonads develop as ovaries (Fig. 1.2b). The mesonephric ducts (Wolffian duct) are responsible for the development of testes, and the paramesonephric ducts are responsible for the development of ovaries.

The paramesonephric (Müllerian) ducts are the precursors of the uterus, fallopian tubes, cervix, and upper vagina. In the 8th week, the paired paramesonephric ducts lie medial to the mesonephric ducts. The paramesonephric ducts fuse to form a confluence. This process is the initial stage in the development of the upper two-thirds of the vagina, the cervix, uterus, and both fallopian tubes (Fig. 1.3a). The medial portions of the paramesonephric ducts fuse to form the uterus and upper vagina (lateral fusion 9–11 weeks), the lateral portions give rise to fallopian tubes. Müllerian organogenesis is complete by 5 months with uterine septal resorption.

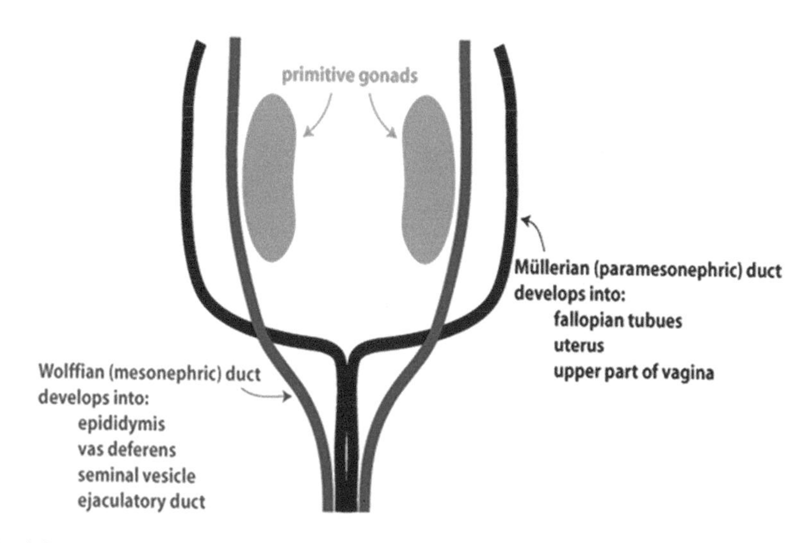

Fig. 1.1 Indifferentiated gonad. Indifferent phase of gonads with both Wolffian (mesonephric) and Müllerian (paramesonephric) ducts present

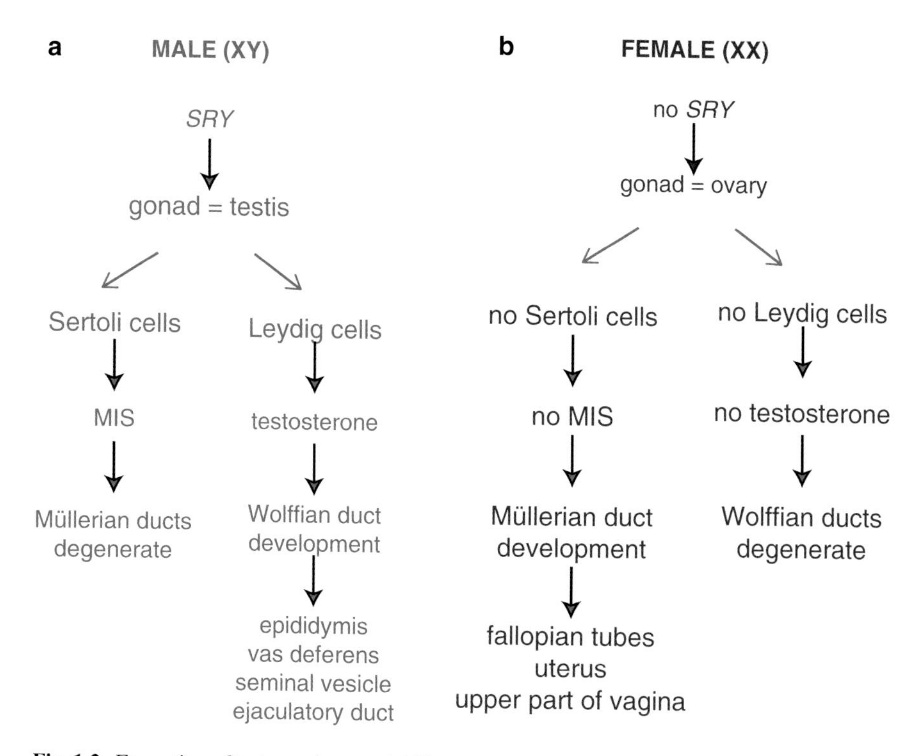

Fig. 1.2 Formation of testes and ovary. (**a**) The Wolffian duct becomes testes with the presence of SRY gene. (**b**) The Müllerian duct becomes ovary in the absence of SRY gene

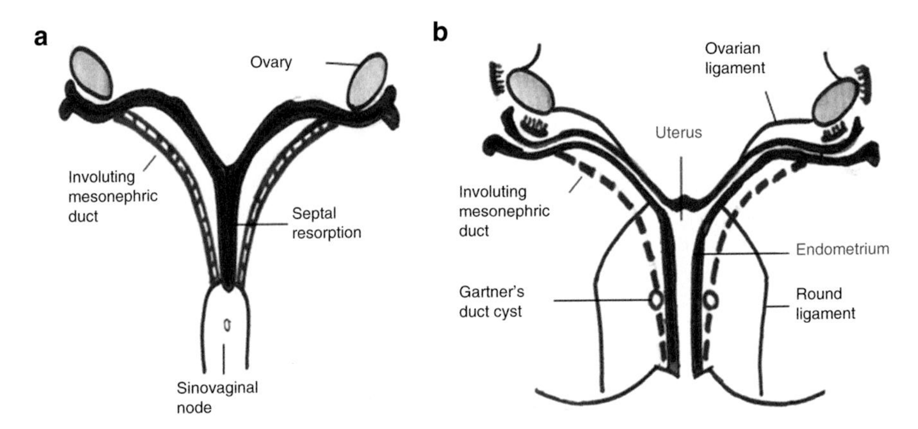

Fig. 1.3 Development of uterus. At 7 weeks, the paramesonephric ducts differentiate while the mesonephric ducts involute. Müllerian organogenesis is complete by 5 months with uterine septal resorption (**a**); the medial portions of the paramesonephric ducts fuse to form the uterus and upper vagina (lateral fusion 9–11 weeks), the lateral portions give rise to fallopian tubes. The female internal genitalia at 5 months (**b**)

The endometrium differentiation is complete by the 20th week, which is derived from the mucosal lining of the fused Müllerian ducts (Fig. 1.3b). By the 22nd week, the canalization is complete with the formation of the uterine cavity, cervical canal, and the vagina.

1.1.2 *Tissue Layers of the Uterus*

The walls of the uterus are composed mainly of smooth muscle fibers and fibrous tissues. They consist of three layers: (a) endometrium or mucosal lining; (b) myometrium, a smooth muscular layer; and (c) perimetrium, the peritoneal lining outside the wall. From puberty until menopause, the endometrium undergoes cyclical changes under the control of ovarian steroid hormones. Some elastic tissue, blood vessels, nerves and lymphatics are also present.

- The *perimetrium* is the outermost layer that forms the external skin of the uterus. It is continuous with the peritoneum that covers the major organs of the abdomino pelvic cavity. The perimetrium protects the uterus from friction in two ways; by forming a smooth layer of simple squamous epithelium and by secreting watery serous fluid to lubricate its surface.
- The *myometrium* forms the middle layer of the uterus and contains many layers of visceral muscle tissue. Expansion and contraction of the uterus during pregnancy is because of myometrium.

- Inside the myometrium is the *endometrium* layer that lines the uterus. It is made of simple columnar epithelium with glands resting on a layer of connective tissue, the stroma. This highly vascular connective tissue provides support to the developing embryo and fetus during pregnancy.

Around the time of ovulation, the uterus builds a thick layer of vascular endometrial tissue in preparation to receive a zygote. If fertilization fails, it would trigger the blood vessels of the endometrium to atrophy and the uterine lining to be shed. This causes menstruation approximately every 28 days for most women. In the case of successful fertilization, a zygote will implant itself into the endometrial lining, where it begins to develop to form a fetus. The uterus plays a critical role in the process of childbirth. Prior to delivery, hormones trigger waves of smooth muscle contraction in the myometrium that slowly increase in strength and frequency. At the same time, the smooth muscle tissue of the cervix begins to efface, or thin, and dilates. Once the cervix is fully dilated, the uterine contractions drastically increase in intensity and duration until the fetus comes out.

Further Reading

Andrew Haley 2012 from G. Mann et al. (eds.), Imaging of Gynecological Disorders in Infants and Children, Medical Radiology. Diagnostic Imaging, DOI: 10.1007/174_2010_128, _Springer-Verlag Berlin Heidelberg 2012.

Moore KL, Persaud TVN (1998) The urogenital system. In: Moore KL, Persaud TVN (eds) The developing human. Clinically oriented embryology, ed 6. WB Saunders, Philadelphia, pp 303–347.

Spencer TE, Hayashi K, Hu J et al., (2005). Comparative Developmental Biology of the mammalian uterus. *Current top developmental biology* 68, 85–122.

Susan Wray. (2007). “Insights into the uterus”. Experimental Physiology 92.4, 621.

1.2 Endometrium: Inner Lining of the Uterus

Human endometrium is the inner mucosal layer that lines the uterine cavity as far the isthmus of the uterus, where it becomes continuous with the lining of the cervical canal. The endometrium is opposed to the outer perimetrium and median myometrium. There is no submucosal tissue to separate endometrial glandular tissue from underlying smooth muscle. Both endometrium and myometrium originate from the Müllerian ducts during embryonic life. Endometrium is a simple columnar epithelium (tissue that covers all free body surfaces), resting on a layer of connective tissue, the stroma. Tubular glands reach through the endometrial surface to the base of stroma.

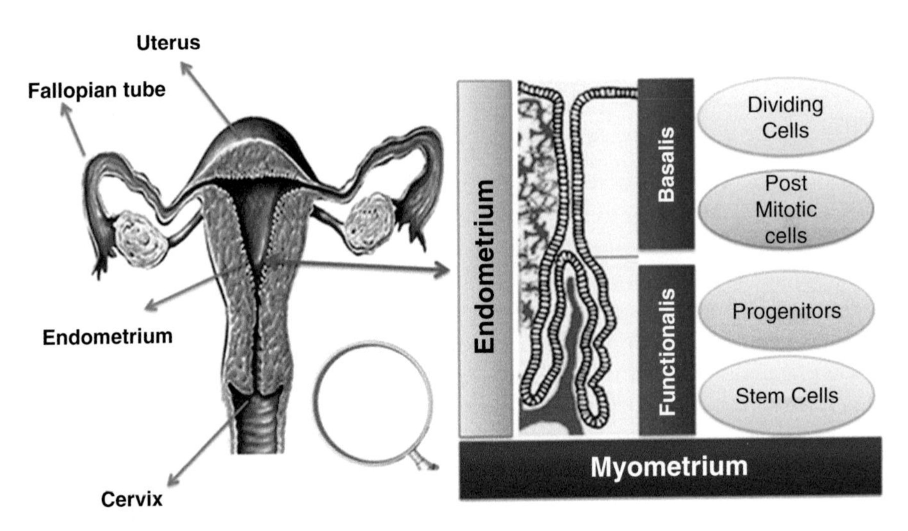

Fig. 1.4 Layers of endometrium. The superficial layer (functionalis) is shed during menstruation, whereas the permanent bottom layer (basalis) gives rise to the regeneration of endometrium after each menstruation. Cells of the layers are shown

1.2.1 Endometrial Cell Types

In a woman of reproductive age, the endometrium itself is structurally and functionally divided into two relatively distinct layers: the "Stratum Functionalis" layer, the upper two-thirds, and the "stratum basalis" layer, the lower one-third (Fig. 1.4). It is the functionalis layer that undergoes cyclical degeneration and proliferation in response to ovarian hormones and is destined to adequately mature for embryo implantation. It sheds during menstruation. Basalis (basal layer), adjacent to myometrium, provides basal region of glands, dense stroma, and lymphoid aggregates. It forms new functional layer after menstruation. The cellular components of human endometrium can be primarily divided as follows: the epithelial cells (luminal and glandular) and the supporting mesenchymal cells (stromal cells) as well as vascular (endothelial) cells (Fig. 1.5). Furthermore, the endometrial stroma of both layers is populated by different classes of leukocytes, including the tissue-specific uterine natural killer cells, mast cells, macrophages, T and B cells, and neutrophils.

1.2.1.1 Epithelium

This layer is interconnected with specialized regions of plasma membrane containing cell junctions and is divided into luminal (surface or superficial) epithelium and glandular epithelium submerged into the stromal compartment. The luminal epithelium covers the endometrial surface, functioning as a primary barrier of defense against infections, as well as serving as a surface for the attachment of the blastocyst

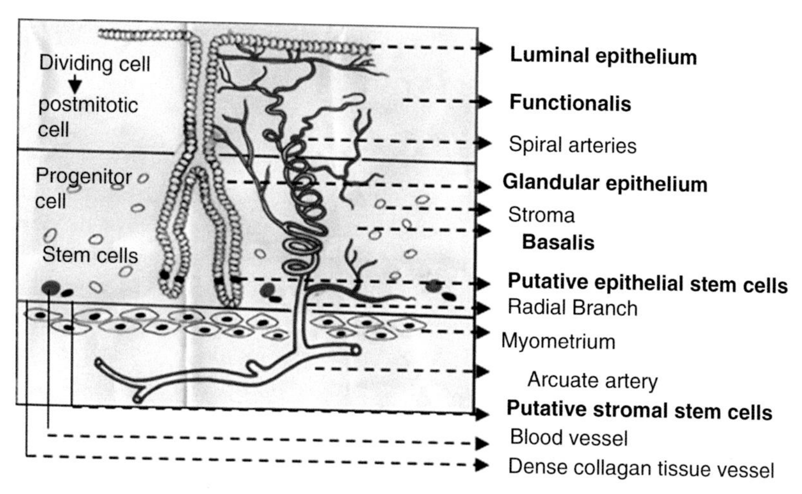

Fig. 1.5 A single, continuous layer of epithelial cells lines the surface of the stroma and penetrates the stroma. The entire thickness of the endometrium is penetrated by the spiral arteries and their capillaries. The striking changes in the spiral arteries are consistently observed before the onset of every menstruation episode. The putative epithelial and stromal stem cells also shown

during the implantation process. The glandular epithelial cells consist of a single layer of columnar cells with varying morphology, depending on different phase of the menstrual cycle, that secrete glycoproteins, glycogen, and cytokines and express cell adhesion molecules during the secretory phase of the menstrual cycle.

1.2.1.2 Stromal Compartment

The stromal compartment includes mainly stromal cells, leucocytes, macrophages, and endothelial cells that vary in their numbers according to different phases of menstrual cycle. They produce matrix metalloproteinases (MMPs), tissue inhibitor metalloproteinases (TIMPs), vascular endothelial growth factor (VEGF), and many other cytokines that are involved in angiogenesis and recruitment of inflammatory mediators in endometrial shedding during menstrual phase. The stromal cell compartment is generally less well studied than the epithelial compartment. Mitoses of stromal cells reach a maximum (about 10 per 1000 cells) around the time of ovulation, and this is later reduced to almost zero over the peri-implantation period, with an increase when nearing the menstrual phase. The stromal cells prepare for implantation by increasing their mean nuclear profile and stromal cell density from LH+2 to LH+6. During this period, the cells are less densely packed than at any other point of time, with maximal oedema, thus making it possible for the blastocyst to implant. At this same time, there are numerous pre-decidual and pseudo-decidual changes closer to the blood vessels providing the nutritive support by secreting glycogen.

1.2.1.3 Decidual Changes

Once the embryo invades the luminal epithelium, a decidual response is elicited, characterized by transformation of stromal cells into secretory, epithelioid-like decidual cells, and influx of specialized uterine immune cells and vascular remodeling. In humans, this decidual response is primarily under maternal control and initiated in the mid-secretory phase of each menstrual cycle, irrespective of whether pregnancy has occurred or not. In the absence of pregnancy, spontaneous menstruation occurs, which correlates that integrity of decidual endometrium is dependent on progesterone signaling. Thus, in absence of progesterone, the expression of pro-inflammatory cytokines, chemokines, and matrix metalloproteinases will activating a sequence of events and tissue breakdown, resulting in menstruation.

Further Reading

Brosens, J. J., Parker, M. G., McIndoe, A., Pijnenborg, R. and Brosens, I. A. (2009) A role for menstruation in preconditioning the uterus for successful pregnancy. *Am J Obstet Gynecol*, **200**, 615 e611–616.

Brosens, J. J., Pijnenborg, R. and Brosens, I. A. (2002) The myometrial junctional zone spiral arteries in normal and abnormal pregnancies: a review of the literature.*Am J Obstet Gynecol*, **187**, 1416–1423.

Cunningham, F. G., Williams, J. W. and Dawsonera (2010) Williams obstetrics. McGraw-Hill Medical, New York.

Gargett CE. Uterine stem cells: what is the evidence? Hum Reprod Update. 2007;13:87–101.

Jabbour, H. N., Kelly, R. W., Fraser, H. M. and Critchley, H. O. (2006) Endocrine regulation of menstruation. *Endocrine reviews*, **27**, 17–46.

Macklon, N. S., Geraedts, J. P. and Fauser, B. C. (2002) Conception to ongoing pregnancy: the 'black box' of early pregnancy loss. *Hum Reprod Update*, **8**, 333–343.

More, I. A., Armstrong, E. M., Carty, M. and McSeveney, D. (1974) Cyclical changes in the ultrastructure of the normal human endometrial stromal cell. *J Obstet Gynaecol Br Commonw*, **81**, 337–347.

Noyes, R. W., Hertig, A. T. and Rock, J. (1975) Dating the endometrial biopsy.*Am J Obstet Gynecol*, **122**, 262–263.

1.2.2 Endometrial Cycle

For the purpose of periodic elimination of the endometrium that undergoes regression, shrinkage, and necrosis at end of each cycle, the uterus undergoes physiological bleeding. The endometrial cycle, primarily regulated by the hypothalamo-pituitary-ovarian axis, undergoes a sequence of histological events endowed to fulfill its goal "to receive and nurture an embryo." In a failure to do so,

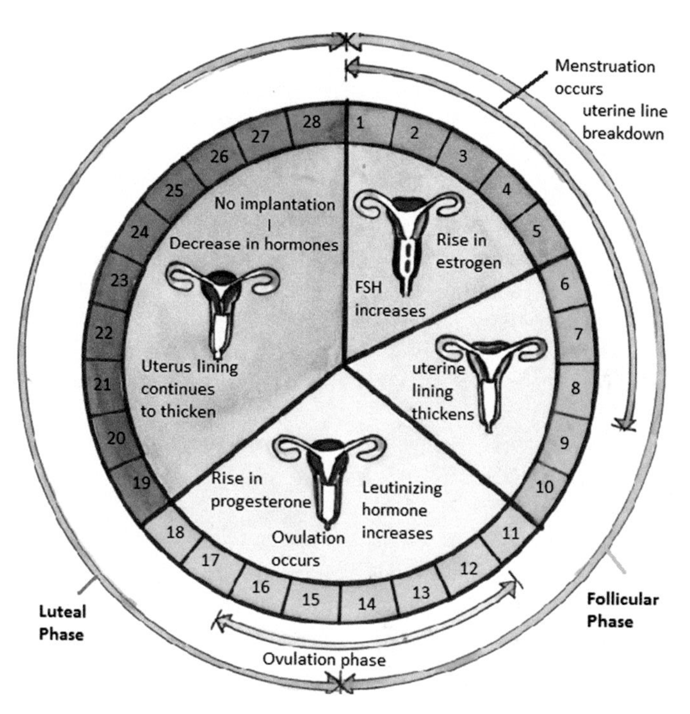

Fig. 1.6 Different phases of menstrual cycle. It shows the different stages of menstruation with its 28-day cycle along with hormonal regulation and ovulation

it succumbs to "the fall – the menstrual phase," undergoing aseptic inflammatory/apoptotic breakdown and shedding. The functionalis layer is shed each month with menstruation and is then regenerated from the basalis layer which is not shed. During this menstrual cycle, the endometrium undergoes three stages: the follicular or proliferative phase, the secretory or progestational phase, and the menstrual phase (Fig. 1.6).

The proliferative phase begins at the end of the menstrual phase, is under the influence of estrogen, and parallels growth of the ovarian follicles. The early proliferative phase is characterized by the presence of narrow glands, thin surface epithelium, and compact stroma. During the mid- and late proliferative phase, the endometrium becomes thick with prominent stroma. The secretory phase begins approximately 2–3 days after ovulation when progesterone is produced by the corpus luteum. The spiral arteries and glands begin to entwine with visible presence of luminal secretions in order to prepare itself for the arrival of the embryo. MMPs and other inflammatory mediators aid this process. Menstrual phase begins with the shedding of the endometrium (compact and spongy layers) when no fertilization occurs. This results in apoptotic and necrotic changes in the endometrial functionalis zone and leads to the expulsion of the functional layer of the endometrium. Menstruation ensues for 3–7 days, as the next ovulatory cycle revives.

Further Reading

Brosens, J. J., Parker, M. G., McIndoe, A., Pijnenborg, R. and Brosens, I. A. (2009) A role for menstruation in preconditioning the uterus for successful pregnancy. *Am J Obstet Gynecol*, 200, 615 e611–616.

Jabbour, H. N., Kelly, R. W., Fraser, H. M. and Critchley, H. O. (2006) Endocrine regulation of menstruation. *Endocrine reviews*, 27, 17–46.

King, A. E. and Critchley, H. O. (2010) Oestrogen and progesterone regulation of inflammatory processes in the human endometrium. *The Journal of steroid biochemistry and molecular biology*, 120, 116 126.

Macklon, N. S., Geraedts, J. P. and Fauser, B. C. (2002) Conception to ongoing pregnancy: the 'black box' of early pregnancy loss. *Hum Reprod Update*, 8, 333–343.

McLennan CE, Rydell AH. Extent of endometrial shedding during normal menstruation. Obstet Gynecol. 1965;26:605–21.

More, I. A., Armstrong, E. M., Carty, M. and McSeveney, D. (1974) Cyclical changes in the ultrastructure of the normal human endometrial stromal cell. *J Obstet Gynaecol Br Commonw*, 81, 337–347.

Padykula HA, Coles LG, McCracken JA, et al. A zonal pattern of cell proliferation and differentiation in the rhesus endometrium during the estrogen surge. Biol Reprod. 1984;31(5):1103–18.

Schwab KE, Chan RW, Gargett CE (2005). Putative stem cell activity of human endometrial epithelial and stromal cells during the menstrual cycle. *Fertil Steril* 84(Suppl.2), 1124–1130.

Sherman B, Korenman S. Hormonal characteristics of the human menstrual cycle throughout reproductive life. J Clin Invest. 1975;55:699–706.

1.3 Endometrial Stem Cells: An Overview

The human endometrium is a dramatic tissue that grows about 7 mm within 1 week in every menstrual cycle. At this rate, there is a very rapid rate of angiogenesis for approximately 400 cycles within a tightly controlled manner. This justifies the significance of endometrial stem cells, as the rapid growth is accompanied by these stem cells. Endometrium is divided into two zones: the inner functionalis, which is adjacent to the uterine cavity, and a deeper basalis layer, which overlies the myometrium. The functionalis layer is shed each month with menstruation and is then regenerated from the basalis layer which is not shed. The functionalis, comprising the upper two-third of the endometrium, is divided into stratum compactum and stratum spongiosum. The stratum compactum is a superficial thin layer nearest to the uterine cavity and contains the lining cells, necks of the uterine gland, and relatively dense stroma. The stratum spongiosum is the deeper part of functionalis composed of main portions of the uterine glands and accompanying blood vessels; the stromal cells are more loosely arranged and larger than in the stratum compactum.

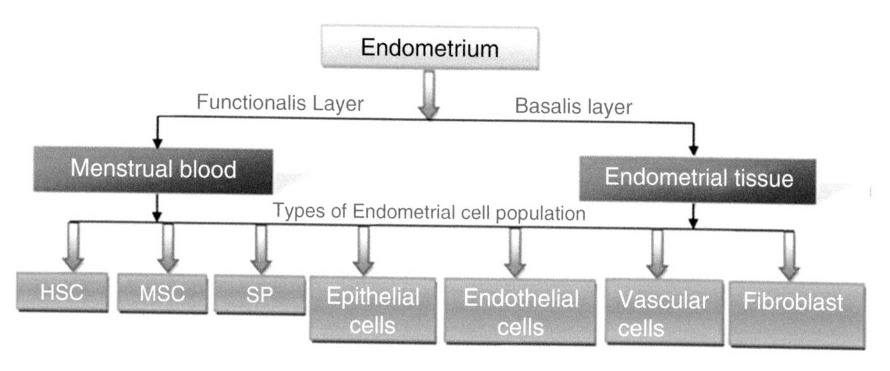

Fig. 1.7 Key cell types in research and therapeutics

The lower basalis contains the basal region of the uterine glands, dense stroma (that remains relatively unaltered during the menstrual cycle), large blood vessel remains, and lymphoid aggregates. It serves as the germinal compartment for generating new functionalis each month.

It has been postulated that the niche of these adult stem or progenitor cells of the endometrium is the lower basalis. These stem or progenitor cells were also identified to be in the trophic endometrium of post-menopausal women. The main cell populations present in the functional stratus are epithelial cells, stromal cells, and leukocytes. Epithelial cells cover the luminal surface and tubular glands in basal and functional layers. Endometrial connective tissue, the stroma, comprised of fibroblasts that rapidly differentiate into decidualized cells when stimulated by an implanting blastocyst. The stromal compartment also compasses lymphocytes, granulocytes, and macrophages during luteal phase of the menstrual cycle. Epithelial and stromal cells are the main target of paracrine signals of proliferation and differentiation. During normal menstrual cycles, human endometrium undergoes cyclic construction and sloughing. The functionalis is lost while the basal layer containing the deep glandular epithelium gets preserved. Later on, the functionalis develops from the stem cells of basalis layer in response to the appropriate hormonal stimulus, regenerating the whole endometrium. The key cell types of endometrial tissue that has gained significant attention in research and therapeutics are listed (Fig 1.7).

It is, hence, evident that tissue remodeling occurs during each menstrual cycle, after resection, parturition, and in post-menopausal women using hormone replacement therapy. This dynamic remodeling and regeneration during the menstrual cycle and pregnancy has been controlled by the adult stem cells of the uterus. Besides, bone marrow-derived stem/progenitor cells also contribute to the vascular remodeling or transdifferentiating into endometrial cells. In parallel, emerging research results suggest that irregular function of these stem cells may contribute to the abnormalities substantially contributing to endometrial disorders, including endometriosis, causing dysmenorrhea, adenomyosis, subfertility, endometrial hyperplasia, and endometrial carcinoma. Such stem cells may also contribute to the pathogenetic process, because their high proliferation promotes rapid clonal expansion. Thus, I call endometrial stem cells as a "double edged sword"; wherein, on one side, it plays a dynamic role in normal physiology and development throughout women's life as well, on the other side, it contributes to the pathogenesis of various endometrial disorders through its abnormal proliferation and differentiation.

Thus, reviewing the currently available data regarding the stem cell involvement in normal uterine tissue maintenance and development, and its role in causing various endometrial disorders, giving particular attention to the potential clinical application of the knowledge stemming from basic endometrial science research is most imperative.

Further Reading

Gargett CE. Uterine stem cells: what is the evidence? Hum Reprod Update. 2007;13:87–101.

Gargett CE, Masuda H: Adult stem cells in the endometrium. Mol Hum Reprod 2010, 16(11):818–834.

McLennan CE, Rydell AH. Extent of endometrial shedding during normal menstruation. Obstet Gynecol. 1965;26:605–21

Padykula HA, Coles LG, McCracken JA, et al. A zonal pattern of cell proliferation and differentiation in the rhesus endometrium during the estrogen surge. Biol Reprod. 1984;31(5):1103–18.

Padykula HA: Regeneration in the primate uterus: the role of stem cells. Ann NY Acad Sci 1991, 622:47–56.

Sasson IE, Taylor HS: Stem cells and the pathogenesis of endometriosis. Ann NY Acad Sci 2008, 1127(1):106–115.

Schwab KE, Chan RW, Gargett CE (2005). Putative stem cell activity of human endometrial epithelial and stromal cells during the menstrual cycle. *Fertil Steril* 84(Suppl.2), 1124–1130.

Spencer TE, Hayashi K, Hu J et al., (2005). Comparative Developmental Biology of the mammalian uterus. Current top developmental biology 68, 85–122.

1.3.1 *Identification of Endometrial Stem Cells*

Despite the advances in recent years in the isolation and applicability of stem cells from many sources, there exist only limited studies that consider the availability and applicability of stem cells in female reproductive organs. However, studies of adult stem cell biology in the uterus is growing in recent years with the fact that the uterus undergoes perhaps the most extensive proliferative changes and remodeling in adult mammals compared with other organs. It has been postulated that the niche of these adult stem or progenitor cells of the endometrium is the lower basalis. Adult stem cells in the endometrium were difficult to identify in the past due to lack of precisely characterized cell surface markers specific for adult stem cells of the endometrium.

Later, Studies have provided an indirect evidence for the existence of endometrial stem cells by characterizing cell populations in the endometrium, which exhibit the functional properties of stem cells. Recent evidence from the literature on the existence of epithelial and stromal/stem cells in endometrium has been identified, and research regarding its potent applications is underway. The idea that the endometrium contains stem or progenitor cells was demonstrated first by Prianishnikov in the year 1978. Following this, the majority of studies on its unique properties and its potential applications have been proposed several times over the past few decades.

Clonogenicity, as defined by the ability of single cells to proliferate and produce isolated colonies of progeny when seeded in culture at very low density, is the initial step to identify stem or progenitor cells. Based on this property, the first published direct evidence for the existence of adult stem/progenitor cells in the human uterus identified in the endometrium is clonogenic epithelial and stromal cells, suggesting the presence of two types of adult stem/progenitor cells. Using a purified single-cell suspension dispersed from hysterectomy specimens, Chan et al. identified a small population of stromal cells (1.25 %) and epithelial (0.22 %) cells in human endometrium that possessed clonogenic activity. The ability of large epithelial CFU to self-renew and differentiate identified epithelial progenitor cells, whereas small CFU are identified to be mature transit-amplifying cells. It was demonstrated that these stem/progenitor cells reside in basalis and are responsible for regeneration of functionalis layer. The mature transit-amplifying cells reside in functionalis that are responsible for epithelial proliferation in the first half of the menstrual cycle. Clonogenicity was similar between actively and inactively menstruating women, suggesting that these clonogenic cells are not controlled by ovarian-derived steroid hormones. Based on in vitro studies, growth factors supporting clonogenic cells include epidermal growth factor (EGF), platelet-derived growth factor BB (PDGF-BB), and transforming growth factor (TGF) α.

The LRC approach was used to identify adult stem cells when specific markers were unknown. Epithelial LRCs did not express estrogen receptor-alpha (ER-alpha). Although most of the stromal LRCs were estrogen receptor (ER) negative, 16 % of the stromal LRCs were found to express ER. This minor population represented a unique population of estrogen responsive stem/progenitor cells and found to transmit paracrine signals to epithelial cells for endometrial epithelium regeneration. In ovariectomized pre-pubertal mice, the first cells to proliferate in estrogen-stimulated endometrial growth are the epithelial LRC, suggesting that they function as stem/progenitor cells to initiate epithelial regeneration. However, in ovariectomized cycling mice, both epithelial LRC and non-LRC rapidly proliferated in response to estrogen to regenerate luminal and glandular epithelium. Thus, this study demonstrated for the first time the presence of both epithelial and stromal LRCs in mouse endometrium, suggesting that these stem-like cells may be responsible for endometrial regeneration.

Some studies of endometrial stem cells have focused on the existence of an extrauterine source of adult stem cells such as bone marrow that are presumably recruited to the uterus with each menstrual/estrous cycle, and that it is these cells that are responsible for repair of the uterus following menses or parturition.

Independent data generated from various laboratories suggest a role for bone marrow contributions to the uterine endometrium. In a human study carried out by Taylor group, uterine tissues were obtained from four women following hysterectomy who previously had received a single bone marrow transplant following chemotherapeutic conditioning, and donor-derived cells were found in both the epithelial and stromal compartments of the endometrium. In a subsequent study completed by the same laboratory, male bone marrow-derived cells were found in the uterine endometrium of female mice by transplanting male bone marrow cells to female mice by lethal irradiation. Later, another report carried out by Bratincsak et al. confirmed that bone marrow progenitor cells contribute to the uterine epithelium, and the population of cells may include CD45+ cells. Y chromosome fluorescence in situ hybridization was used to detect male bone marrow-derived cells in both the stromal and epithelial compartments. Furthermore, in 2008, a new report by Mints et al. showed that bone marrow-derived endothelial progenitors contribute to the formation of new blood vessels in the endometrium. This again proves that bone marrow acts as an extrauterine source of stem cells that are recruited to the uterus, to repopulate the endometrium.

The characterization of endometrial stem cells was also done on the basis of Hoechst 33342 dye exclusion through ATP-binding cassette transporters. This is called the side population-based approach, as those cells are characterized by their ability to exclude the DNA binding dye Hoechst 33343 by expressing ATP Binding cassette transporter proteins during fluorescence-activated cell sorter (FACS) analysis. Kato et al. identified 0.00–5.11 % as SP cells in normal human endometrium. The clonogenic potential of these cells is yet to be determined. However, SP cells identified from human endometrium display long-term proliferative properties as well as multi-potent differentiation ability. This includes osteocytes, adipocytes, chondrocytes, and also endometrial glandular, epithelial, stromal, and endothelial cells. Thus, it is evident that SP cells are indeed adult stem cells of the endometrium.

Further, Schwab and Gargett also demonstrated the existence of endometrial stem cell through characterization of perivascular markers CD146 and CD140b (PDGF-Rβ). These perivascular markers enabled isolation of stromal cells from human endometrium, which exhibit phenotypic and functional properties of MSC. This finding is supported by more extensive differentiation studies in which rare (1.5 %) CD146+/PDGF-Rβ+ stromal cells could be induced to differentiate into osteocytes, chondrocytes, myocytes, and adipocytes.

Further Reading

Cervello I, Martinez-Conejero JA, Horcajadas JA, Pellicer A, Simon C. Identification, characterization and co-localization of label-retaining cell population in mouse endometrium with typical undifferentiated markers. Hum Reprod. 2007;22:45–51.

Chan RW, Gargett CE. Identification of label-retaining cells in mouse endometrium. Stem Cells. 2006;24:1529–38.

Chan RWS, Schwab KE, Gargett CE (2004). Clonogenicity of Human Endometrial Epithelial and Stromal Cells. *Biology of Reproduction* 70,1738–1750.
Du H, Taylor HS. Contribution of bone marrow-derived stem cells to endometrium and endometriosis. Stem Cells. 2007;25:2082–6.
Gargett CE, Chan RWS, Schwab KE (2007). Endometrial stem cells. Reproductive endocrinology 19(4), 377–383.
Gargett CE, Masuda H (2010). Adult stem cells in the endometrium. *Molecular Human Reproduction* 16(11), 818–834.
Gargett. CE (2006). Identification and characterization of human endometrial stem/ progenitor cells. *Aust N Z J Obstet Gynaecol* 46, 250–253.
Goodell MA, Brose K, Paradis G et al., (1996). Isolation and functional properties of murine hematopoietic stem cells that are replicating in vivo. *Journal of Experimental Medicine* 183(40), 1797–1806.
I Cerevello, A Mas, C Gil-Sanchis et al., (2011). Reconstruction of endometrium from endometrial side population cells lines. *Plos one* 6(6).
Kato K, Yoshimoto M, Kato K, Adachi S, Yamayoshi A, Arima T, Asanoma K, Kyo S, Nakahata T, Wake N. Characterization of sidepopulation cells in human normal endometrium. Hum Reprod. 2007;22:1214–23.
Masuda H, Matsuzaki Y, Hiratsu E et al., (2010). Stem cell- Like properties of the Endometrial side population: Implication in Endometrial regeneration. *Plos one* 5(4).
Mints M, Jansson M, Sadeghi B, et al. Endometrial endothelial cells are derived from donor stem cells in a bone marrow transplant recipient. Hum Reprod. 2008; 23:139–143
Ono M, Maruyama T, Masuda H, Kajitani T, Nagashima T, Arase T, Ito M, Ohta K, Uchida H, Asada H, et al. Side population in human uterine myometrium displays phenotypic and functional characteristics of myometrial stem cells. Proc Natl Acad Sci U S A. 2007;104: 18700–5.
Schwab KE, Chan RW, Gargett CE (2005). Putative stem cell activity of human endometrial epithelial and stromal cells during the menstrual cycle. *Fertil Steril* 84(Suppl.2), 1124–1130.
Taylor HS. Endometrial cells derived from donor stem cells in bone marrow transplant recipients. JAMA. 2004;292:81–5.
Tsuji S, Yoshimoto M, Kato K et al., (2008). Side population cells contribute to the genesis of human endometrium. *Fertil Steril* 90, 1528–1537.

1.4 Stem Cells from Menstrual Blood

To date, stem cells have been found to be obtained from the uterus in a variety of ways including hysterectomy, diagnostic curettage, menstrual blood, and first trimester deciduas. It is obvious that human endometrium shows a strong ability to regenerate, but the cellular sources that are responsible are still not understood.

Thus, it is important to study the nature of these cells in order to reveal their functions. There are three distinct endometrial stem cells including epithelial progenitor cells, mesenchymal stem cells, and endothelial progenitor cells. Menstrual blood has become the convenient source due to the ease of availability, its plasticity, and longevity of the cells. The stem cells derived from menstrual blood are reported to provide great promises for use in tissue repair and treatment of diseases. A study published in 2007 was the first to identify and characterize a new source of stem cells within menstrual fluid. It showed that menstrual-derived stem cells (MenSCs) are rapidly expanded and differentiated under standard laboratory conditions. The rationale of why people study endometrial stem cells from menstrual blood is as follows:

Human endometrium is unique in its temporally regulated processes of cellular proliferation, differentiation, and shedding of the functionalis layer with each menstrual cycle. The menstrual cycle which is required for reproduction is controlled by the endocrine system. During this period of time, the uterus experiences various cyclical changes of endometrial thickening, vascular proliferation, glandular secretion, and endometrial growth followed by the shedding of the functionalis of the endometrium. With presence of progesterone, endometrium prepares itself for implantation. However, absence of progesterone, the demise of corpus luteum, and the subsequent fall in circulating progesterone lead to vasoconstriction, necrosis of the endometrium, and menstruation. Thus, there is growing interest in their clinical potential on menstrual blood since they display a high proliferation rate that supports the uterus. These multi-potent cells are obtainable in periodic in a non-invasive manner, devoid of the biological and ethical issues concerning other stem cell types. For these reasons, reliable studies on menstrual blood-derived stem cells are in process.

For example, menstrual blood has been studied for its viability and cell characteristics with some initial work published in 2008 in Cell Transplantation. Menstrual blood-derived stem cells had been demonstrated to possess characteristics of self-renewal, high proliferative potential in vitro, and the ability to differentiate toward diverse cell lineages in the induction media, which was demonstrated through the in vitro and in vivo studies of characterization, proliferation, and differentiation. These cells are called endometrial regenerative cells (ERCs). They are found to be heterogenic and vary in its morphological phenotype. Epithelial cells are not identified in menstrual blood, but stromal cells have been cultured from menstrual blood. This suggests that only stromal cells are shed during menstruation but not epithelial cells. During the proliferative menstrual phase, the endometrium prepares for recruiting new endometrial stem cells to cover the new blood vessels. The variation in the numbers of endothelial progenitor cells and stem cell-derived factor-1 is a pivotal factor for release and homing of stem cells. However, the rationale for the same on the identification and release of stromal and epithelial cells in menstrual blood and its mechanism needs to be further investigated.

Although menstrual blood stem cell research are underway, adult stem or progenitor cells that are responsible for the cyclical regeneration of the endometrium functionalis reside in the basalis region of the endometrium. Hence, the study of these stem cells from the basalis layer of the endometrial tissue is of utmost important. Based on the dynamic tissue remodeling in all compartments of the uterus, during the menstrual cycle and pregnancy, direct studies on basalis region are also underway. Hence, a thorough characterization of the uterine/endometrial stem cells derived from the endometrial tissue biopsy of the inner lining of the uterus or from the intact uterus surgically removed through hysterectomy is equally important/more important as that of studies on menstrual blood stem cells. This is possible by performing diagnosis of abnormal uterine bleeding and can be done easily in an office setting with a small curette or the self-contained small vacuum cannula (Pipelle™), which minimizes the patient's discomfort. Besides, it is reported that irregular function of these stem cells only contribute to the pathogenesis of various gynecological disorders such as endometriosis, endometrial hyperplasia, infertility, endometrial cancer, and so on. Thus, studying the potential of the so-called actual stem cells that are residing in the basalis layer of the endometrium is imperative.

Once a mechanical or functional characteristic platform of these endometrial stem cells has been constructed, it then becomes easier to understand/implement the following parameters:

1. The underlying role of these endometrial stem cells in the morphogenesis and normal physiological development of the uterus/endometrium throughout women's life
2. The underlying role of these endometrial stem cells in causing pathophysiology of various gynecological disorders, as specified above
3. To utilize endometrium, the trash source obtained from the uterus, for the treatment of wide horizon of diseases in two different approaches
   (a) As a treasure in regenerative medicine, whereby enhancing the applicability of endometrial stem cells, the highly regenerative cells for treating wide horizon of diseases
   (b) To design possible mechanism to target endometrial stem cells, thereby treating various gynecological disorders such as endometrial hyperplasia, endometrial carcinoma, and so on

Further Reading

Allickson JG, Sanchez A, Yefi menko N, Borlongan CV, Sanberg PR. Recent studies assessing the proliferative capability of a novel adult stem cell identified in menstrual blood. Open Stem Cell J. 2011;3:4–10.

Borlongan CV, Kaneko Y, Maki M, Yu S-J, Ali M, Allickson JG, Sanberg CD, Kuzmin-Nichols N, Sanberg PR: Menstrual blood cells display stem cell-like phenotypic markers and exert neuroprotection following transplantation in experimental stroke. Stem Cells Dev 2010, 19(4):439–452.

Gargett CE, Masuda H: Adult stem cells in the endometrium. Mol Hum Reprod 2010, 16(11):818–834. Robb A, Mills N, Smith I, Short A, Tura-Ceide O, Barclay G, Blomberg A, Critchley H, Newby D, Denison F: Influence of menstrual cycle on circulating endothelial progenitor cells. Hum Reprod 2009, 24(3):619–625.

Hida N, Nishiyama N, Miyoshi S, Kira S, Segawa K, Uyama T, Mori T, Miyado K, Ikegami Y, Cui C: Novel Cardiac Precursor-Like Cells from Human Menstrual Blood Derived Mesenchymal Cells. Stem Cells 2008, 26(7):1695–1704.

Musina R, Belyavski A, Tarusova O, Solovyova E, Sukhikh G: Endometrial mesenchymal stem cells isolated from the menstrual blood. Bull Exp Biol Med 2008, 145(4):539–543.

Patel AN, Park E, Kuzman M, et al. Multipotent menstrual blood stromal stem cells, isolation, characterization and differentiation. Cell Transplant. 2008;17:303–11.

Schüring AN, Schulte N, Kelsch R, Röpke A, Kiesel L, Götte M: Characterization of endometrial mesenchymal stem-like cells obtained by endometrial biopsy during routine diagnostics. Fertil Steril 2011, 95(1):423–426.

Schwab KE, Chan RW, Gargett CE. Putative stem cell activity of human endometrial epithelial and stromal cells during the menstrual cycle. Fertil Steril. 2005;84 Suppl 2:1124–30.

Xiaolong Meng, Thomas E Ichim, Jie Zhong et al., (2007). Endometrial regenerative cells: A novel stem cell population. *J Transl Med.* 5, 57.

Chapter 2
Isolation, Characterization, and Differentiation of Endometrial Stem Cells

2.1 Isolation and Culturing of Endometrial Stem Cells from Menstrual Blood

Isolation and culturing of menstrual blood-derived stem cells was procured by various methods as mentioned below. Although there are variations in collection, isolation, and processing of MenScs according to various researchers, the overall phenomenon remains the same. Menstrual effluent was procured by the use of a sterile menstrual cup, the Divacup, that may be a silicone medical grade menstrual cup. Since we are aware most menstrual cell specimens will have a bioburden level, they are treated with antibiotics and kept cold. The medium used to isolate and culture and the antibiotics used by various researchers vary according to their protocol and the experimental set up. To collect the effluent, the menstrual cup is inserted into the vagina similarly to the process of inserting a tampon. The collection is mostly scheduled during the heaviest flow of the cycle, and the cup will remain in place for no more than 2–4 h, again according to the experimental set up. The volume collected may vary from 5 ml–15 ml. Once the menstrual effluents are procured, the menstrual cup is removed and the blood is transferred to the sterile media/PBS with antibiotics. Optimal transport after collection would be 24 h. The supernatant were evaluated for bacteria. The cells were washed twice and seeded in a culture flask. These cells displayed stromal cell morphology in the serum-containing growth medium in a flask and doubled in number every 24 h. The cells were grown in Chang's (Irvine Scientific, Santa Ana, CA) Complete Media. Chang's media for adhesion and proliferation were evidenced to be effective. Cells were assessed for time to adherence to flask, growth rate, and number of passages. Cells were subcultured by using TrypLE Express (Invitrogen, Carlsbad, CA), washed, and replated in complete media. After the cells were processed, they were tested for characterization.

Meng et al. isolated the mononuclear cells using Ficoll-Histopaque method. After processing, the cell pellet was cultured in DMEM with 20 % FCS, adherent

I. Somasundaram, *Endometrial Stem Cells and Its Potential Applications*,
SpringerBriefs in Stem Cells, DOI 10.1007/978-81-322-2746-5_2

cells were observed after 2 weeks, and cells were passaged and characterized for further study. On a contrary, Hida et al. cultured the isolated cells in DMEM high glucose with 10 % FBS, and Nikoo et al. cultured the cells in DMEM-F12 with 20 % antibiotics. In his study, the cells were subjected to adenoviral infection for differentiation studies. Phuc et al. also isolated the cells using Ficoll-Paque method; however, the obtained cells were subjected to FACS-based sorting for further characterization. Allickson et al. isolated and cultured the cells in Chang's complete medium, and the cultured population were used for CD117+ sorting using magnetic activated cell sorter (MACS). The remaining cells were cryopreserved using precooled DMSO, 30 % PBS, and 10 % HSA.

Many hysterectomies are performed yearly in the country, and thus endometrium is supposed to be a potential source of stem cells. Studies have indicated that these stem cells reside in the superficial layers accessible by endometrial biopsy. As routinely performed for gynecologic purposes and does not impair endometrial function, endometrial stem cells become ideal source. We postulate that endometrial tissue specimens from women who are not receiving hormonal therapy are suitable for isolation of stem cells as compared to menstrual blood, for two reasons that: (a) The actual stem cells of the endometrium reside in the basalis layer of the endometrium, and not the functionalis, which sheds during menstruation; and (b) several endometrial hysterectomies are performed yearly; it is a biological waste, easily accessible, and processed, and hence this trash source will serve as more reliable source of endometrial stem cells as a treasure in regenerative applicability.

Further Reading

Chang-Hao Cui, Taro Uyama, Kenji Miyado, Masanori Terai, Satoru Kyo, Tohru Kiyono, Akihiro Umezawa (2007). Menstrual Blood-derived cells confer human dystrophin expression in the murine model of duchenne molecular dystrophy via cell fusion and myogenic transdifferentiation. *Mol Biol Cell* **18**, 1586–1594

Dincer S. Collection of hemopoietic stem cells in allogeneic female donors during menstrual bleeding. Transfus Apher Sci. 2004;30:175–6.

Hida N, Nishiyama N, Miyoshi S, Kira S, Segawa K, Uyama T, Mori T, Miyado K, Ikegami Y, Cui C, Kiyono T, Kyo S, Shimizu T, Okano T, Sakamoto M, Ogawa S, Umezawa A. Novel cardiac precursor- like cells from human menstrual blood-derived mesenchymal cells. Stem Cells. 2008;26:1695–704.

Meng X, Ichim TE, Zhong J, Rogers A, Yin Z, Jackson J, Wang H, Ge W, Bogin V, Chan KW, Thébaud B, Riordan NH. Endometrial regenerative cells: a novel stem cell population. J Transl Med. 2007;5:57.

Nikoo S, Ebtekar M, Jeddi-Tehrani M, Shervin A, Bozorgmehr M, Kazemnejad S, Zarnani AH. Effect of menstrual blood-derived stromal stem cells on proliferative capacity of peripheral blood mononuclear cells in allogeneic mixed lymphocyte reaction. J Obstet Gynaecol Res. 2012;38(5):804–9.

Patel AN, Park E, Kuzman M, Benetti F, Silva FJ, Allickson JG. Multipotent menstrual blood stromal stem cells: isolation, characterization and differentiation. Cell Transplant. 2008;17:303–11.

Phuc PV, Lam DH, Ngoc VB, Thu DT, Nguyet NT, Ngoc PK. Production of functional dendritic cells from menstrual blood – a new dendritic cell source for immune therapy. In Vitro Cell Dev Biol Anim. 2011;47:368–75.

2.2 Isolation and Culture of Endometrial Stem Cells from Endometrial Tissue Biopsy

Endometrium can be collected from reproductively active women undergoing hysterectomy or endometrial biopsy for non-malignant uterine tumors, fibroids, adenomyosis, and uterine prolapse. Endometrial tissue can be collected from the different phases of endometrial cycle including menstrual, secretory, and proliferative phase. Sampling procedures vary with the aim of research, but samples were usually collected from women not undergoing any kind of hormonal therapy. The endometrial biopsy samples containing the endometrial epithelial and stroma cells are collected in HEPES-buffered Dulbecco modified Eagle medium/Hams F-12 supplemented with antibiotic-antimycotic solution (final concentrations: 100 mg/ml penicillin G sodium, 100 mg/ml streptomycin sulfate, 0.25 mg/ml amphotericin B) and fetal bovine/calf serum or Hanks balanced salt solution containing the antibiotics like streptomycin and penicillin.

Isolation of endometrial stem cells involves processing of the finely chopped tissue samples in phosphate-buffered saline devoid of calcium and magnesium ions. The mechanically minced tissue is further digested with DMEM containing type Ia collagenase. Following tissue digestion, epithelial and stromal cells were separated using filtration through a 100 μm and 45 μm strainer. The cells were then centrifuged at 1000 g for 15 min and may or may not undergo ficoll purification. The cells were then washed with PBS several times. The stromal isolates are cultured and propagated in the complete culture medium. Usually, the cells isolated by a general procedure will be heterogeneous, and the types of cells could be distinguished only upon culture. Thus, a homogenous fibroblastic stromal population becomes prominent with spindle-like cell morphology with centrally located nuclei as a monolayer upon culture. The stromal cell suspension obtained is also purified by negative selection using magnetic Dynabeads coated with specific antibodies to remove epithelial cells (BerEP4) and leukocytes (CD45) or by repeated centrifugation to obtain the epithelial cells and the stromal cells as mentioned above. Purified epithelial and stromal cell suspensions were then separated using sorting technique.

These cells were then cultured separately at clonal density (8–20 cells/cm^2, 500–1200 cells/dish) in DMEM/F-12 medium containing 10 % fetal calf serum, 2 Mm glutamine, and antibiotic-antimycotic on fibronectin-coated 100-mm petri dishes in triplicate, and in limiting dilution and incubated at 37 °C in 5 % CO2. The culture medium for epithelial cells was also supplemented with 10 ng/ml epidermal growth factor to promote growth of epithelial clones. Plates were examined twice/week to ensure clones were established from single cells, and individual colonies were

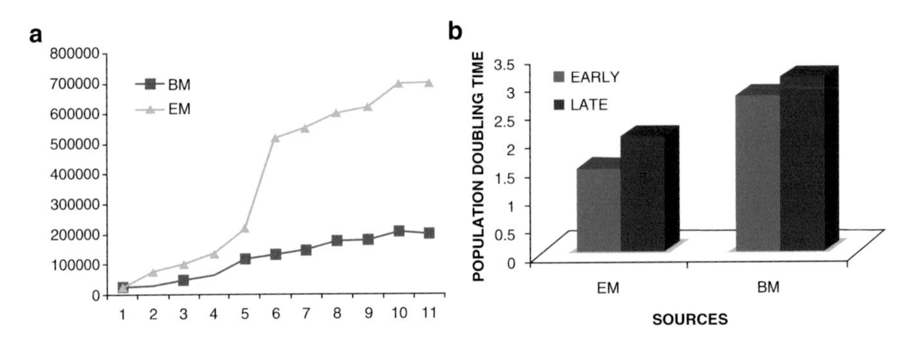

Fig. 2.1 Proliferation of endometrial stem cells. The growth curve and population doubling time of endometrial stem cells in comparison with that of bone marrow. Endometrium shows higher growth potential that could be observed via growth curve (**a**) and PDT (**b**) in comparison with bone marrow

monitored until harvest. Non-overlapping single epithelial or stromal clones were harvested from culture dishes using 0.025 % trypsin and cloning rings after 20–35 days in culture or from individual wells containing a single colony. Six to twelve large clones of epithelial or stromal cells, or 6–12 small clones, were collected from each patient sample for analysis of adult stem cell properties. The self-renewal property of the adult stem cells was monitored using a serial cloning strategy. Large and small clones were obtained, wherein the large clones showed a better self-renewal capacity. Large clones of epithelial and stromal cells were capable of undergoing three rounds of serial cloning.

While most researchers followed the above protocol, some executed a different protocol of obtaining epithelial and stromal cells. The endometrial tissue was collected immediately into Hanks balanced salt solution containing antibiotics like streptomycin (100 mg/ml) and penicillin (100 mg:ml). The tissue was then chopped finely and incubated in Dulbecco's modified Eagle's medium containing 0.1 % (w:v) type 1a collagenase (DMEMC), for 45 min at 37 °C with gentle pipetting every 15 min to disperse the cells. The cell suspension was then centrifuged for 10 min at 100 *g*, leaving a pellet containing epithelial cells and supernatant containing the stromal cells. The epithelial cells were then incubated and pipetted for a further 45 min at 37 °C in DMEMC before repeating the centrifugation. Following the centrifugation, the epithelial cells were re-suspended in 2 ml DMEM containing 2 % glutamine, streptomycin: penicillin (100 mg: ml), and 10 % (v: v) fetal bovine serum. The supernatants containing stromal cells were centrifuged for 5 min at 300 *g*. The resulting supernatant was discarded and the pellet of stromal cells was re-suspended in DMEM. The isolated cells were counted using haemocytometer and re-suspended in complete culture medium. The proliferative potential of endometrium was higher as compared to other post-natal source. This is evident from the figure (Fig. 2.1) that the growth curve and population doubling time of endometrial stem cells in comparison with that of standard bone marrow. The growth potency was sevenfold higher in endometrium as compared to bone marrow; also it showed an increase in population doubling with lesser PDT as compared to bone marrow.

Further Reading

Caroline E. Gargett, Kjiana E. Schwab, Rachel M. Zillwood, Hong P.T. Nguyen and Di Wu (2009). Isolation and Culture of Epithelial Progenitors and Mesenchymal Stem Cells from Human Endometrium. BIOLOGY OF REPRODUCTION 80, 1136–1145.

Chan RW, Schwab KE, Gargett CE. Clonogenicity of human endometrial epithelial and stromal cells. Biol Reprod. 2004;70(6):1738–50.

Chen YJ, Li HY, Chang YL, Yuan CC, Tai LK, Lu KH, Chang CM and Chiou SH (2010) Suppression of migratory/invasive ability and induction of apoptosis in adenomyosis-derived mesenchymal stem cells by cyclooxygenase-2 inhibitors. *FertilSteril*, doi:10.1016/j.fertnstert.2010.01.070.

Chiou SH, Kao CL, Peng CH, Chen SJ, Tarng YW, Ku HH, Chen YC, Shyr YM, Liu RS, Hsu CJ, et al. (2005) A novel in vitro retinal differentiation model by co culturing adult human bone marrow stem cells with retinal pigmented epithelium cells. *Biochem Biophys Res Commun* 326:578–585.

Dimitrov R, Timeva T, Kyurchiev D (2008). Characterization of clonogenic stromal cells isolated from human endometrium. Reproduction 135(4), 551–558.

Dominici M, Le Blanc K, Mueller I, Slaper-Cortenbach I, Marini F, Krause D, Deans R, Keating A, Prockop Dj, Horwitz E (2006). Minimal criteria for defining multipotent mesenchymal stromal cells The International Society for Cellular Therapy position statement. *Cytotherapy* 8(4), 315–317.

Gargett CE, Chan RWS, Schwab KE (2007). Endometrial stem cells. Reproductive endocrinology 19(4), 377–383.

Gargett CE, Masuda H (2010). Adult stem cells in the endometrium. *Molecular Human Reproduction* 16(11), 818–834.

Gargett CE, Schwab KE, Zillwood RM, Nguyen HPT, Wu D (2009). Isolation and Culture of Epithelial Progenitors and Mesenchymal Stem Cells from Human Endometrium. *Biology of Reproduction* 80(6), 1136–1145.

Gargett. CE (2006). Identification and characterization of human endometrial stem/progenitor cells. *Aust N Z J Obstet Gynaecol* 46, 250–253.

Indumathi S, Harikrishnan R, Rajkumar JS, Sudarsanam D and Dhanasekaran M. (2013) Prospective biomarkers of stem cells of human endometrium and fallopian tube in comparison to bone marrow. *Cell Tissue Res* 352(3), 537–549.

Taneera J, Rosengren A, Renstrom E, Nygren JM, Serup P, Rorsman P, and Jacobsen SE (2006) Failure of transplanted bone marrow cells to adopt a pancreatic_-cell fate. *Diabetes* 55:290–296.

Taylor HS. Endometrial cells derived from donor stem cells in bone marrow transplant recipients. JAMA. 2004;292:81–5.

2.3 Phenotypic Characterization

The applicability of stem cells from the human endometrium for regeneration is a fascinating area of research because of the role of these cells in dynamic tissue remodeling and their cyclical regenerative property during the menstrual cycle and pregnancy. Although great breakthrough has been achieved by the identification and isolation of the stem cells from endometrium, the search to identify the biomarkers in human endometrium is still at its infancy. Furthermore, there exist only scanty citations on identification of biomarker expression of the uterine/endometrial stem cells that can isolate/characterize specific cell population. A major advantage of being able to identify the cell surface markers of epithelial and stromal population of the endometrium is that their features can be characterized in non-cultured cells and their utility in cell-based therapies for regenerative medicine could be evaluated in preclinical disease models. Furthermore, a detailed study of endometrial stem cell markers is necessary as pathology of several endometrial disorders such as abnormal uterine bleeding leading to endometrial hyperplasia, infertility, pregnancy complications, miscarriage, endometriosis, and cancer is associated with these endometrial progenitor/stem cells. Thus, the present section focuses at exploring the tissue-specific biomarkers of stem cells derived from the human endometrium. Endometrium exhibits positivity over a wide range of markers, as that of other source of stem cells. In particular, they were demonstrated to exhibit hematopoietic, epithelial, pluripotency, perivascular, mesenchymal, endothelial markers, other cell adhesion molecules and so on, thereby indicating various cell types that could be relevant in tissue repair and regeneration.

Adult stem cells, including MSC, are difficult to purify, since there are no stem cell-specific markers. Various markers or combinations of markers have been used to isolate MSC; however, none was able to isolate a pure MSC population. A positive expression of markers like CD105, CD73, and CD90 and the absence of CD45, CD34, CD14 or CD11b, CD79alpha or CD19, and HLA-DR surface molecules defines mesenchymal stem cells along with its plastic adherent capacity upon culture and multiple differentiation potency, as proposed by the Mesenchymal and Tissue Stem Cell Committee of the International Society for Cellular Therapy (ISCT) criteria that help researchers to identify mesenchymal stem cells and progenitors derived from endometrium as well as their expression prevalence in endometrium-associated diseases. As mentioned here, both MenSCs and EnSCs fulfill these above criteria.

2.3.1 Phenotypic Characterization of Menstrual Blood-Derived Stem Cells

Meng et al. demonstrated the expression of CD9, CD29, CD41a, CD44, CD59, CD73, CD90, CD105, MHC I and the absence of monocyte and hematopoietic stem cell markers on mononuclear cell-derived "endometrial regenerative cells" isolated from menstrual blood. The cells did not express the embryonic stem cell markers SSEA-4 and nanog. Meng et al. also showed that angiogenic factors (ANG-2, vascular endothelial growth factor (VEGF), HGF, and EGF) were constitutively present in the culture media of MenSCs as opposed to cord blood MSC. Allickson also demonstrated CD117+ cell sorting using MACS, as CD117, in combination with the ligand SCF are necessary for the survival and proliferation of germ cells of the testis, the spermatogonia, and the oocytes. He demonstrated the presence of mesenchymal cell markers such as CD13, CD29, CD44, CD49f, CD73, CD90, CD105, CD166, and MHC Class I. In contrast to Meng et al., their study identified pluripotent embryonic stem cell markers SSEA-4, Nanog, and Oct-4. Cesar v. Borlongan et al. also demonstrated the presence of these pluripotent markers along with other aforesaid mesenchymal markers. Other studies also demonstrated various aforesaid markers along with other markers such as CD54, CD10, CD59, CD9, CD41a, CD55, and so on.

Karyotypic analysis demonstrated the maintenance of diploid cells without chromosomal abnormalities. The telomerase activity could be demonstrated comprising about 50 % of the activity of embryonic stem cells, similar to previous findings by Patel et al. Besides, they maintain greater than 50 % of telomerase activity even at passage 12 compared with that in human embryonic stem cells and also appear to mildly express the chemokine receptor CXCR4 and the respective receptor for stromal cell-derived factor-1 (SDF-1), which play a significant role in the mediation of mesenchymal stem cell migration. Zemel'ko et al. demonstrated the presence of CD73, CD90, CD105, CD13, CD29, and CD44. Nikoo et al., showed the presence of CD9, CD29, CD73, CD105, and CD44 and the ES cell marker Oct-4. MenSCs were showed to possess similar marker expression profile as that of bone marrow-derived MSCs by Khanmohammadi and his group with one exception of higher expression of Oct4 in MenScs.

Most recently, study conducted by Franscisca et al. also reported that MenScs expressed mesenchymal stem cells surface markers, such as CD105, CD90, CD73, and CD44, in the absence of hematopoietic cell surface markers including CD34, CD45, and CD14. Furthermore, MenSCs lacked the expression of HLA-DR, CD271, CD117, and the endothelial and epithelial surface marker CD31 and EPCAM in their study. Minimal levels of expression were detected for CD146. Besides, they demonstrated, for the first time, the capacity of MenSCs to support the ex vivo expansion of HSCs, since higher expansion rates of the CD34+CD133+ population as well as higher numbers of early progenitor (CFU-GEMM) colonies were observed.

While most of the markers are commonly present in all studies, there are still differences in the marker expression at least partially due to the variability in donor age, characteristic profile, and variations in isolation procedures. Although common marker profile, which is present in other stem cells, is reported in menstrual blood-derived stem cells, studies on some of the key cell adhesion markers, epithelial markers, and other endothelial markers are still yet to be determined.

Further Reading

Allickson JG, Sanchez A, Yefi menko N, Borlongan CV, Sanberg PR. Recent studies assessing the proliferative capability of a novel adult stem cell identified in menstrual blood. Open Stem Cell J. 2011;3:4–10.

Borlongan CV, Kaneko Y, Maki M, Yu SJ, Ali M, Allickson JG, Sanberg CD, Kuzmin-Nichols N, Sanberg PR. Menstrual blood cells display stem cell-like phenotypic markers and exert neuroprotection following transplantation in experimental stroke. Stem Cells Dev. 2010;19:439–52.

Cui CH, Uyama T, Miyado K, Terai M, Kyo S, Kiyono T, Umezawa A. Menstrual blood-derived cells confer human dystrophin expression in the murine model of Duchenne muscular dystrophy via cell fusion and myogenic transdifferentiation. Mol Biol Cell. 2007;18:1586–94.

Hida N, Nishiyama N, Miyoshi S, Kira S, Segawa K, Uyama T, Mori T, Miyado K, Ikegami Y, Cui C, Kiyono T, Kyo S, Shimizu T, Okano T, Sakamoto M, Ogawa S, Umezawa A. Novel cardiac precursor- like cells from human menstrual blood-derived mesenchymal cells. Stem Cells. 2008;26:1695–704.

Khanmohammadi M, Khanjani S, Bakhtyari MS, Zarnani AH, Edalatkhah H, Akhondi MM, Mirzadegan E, Kamali K, Alimoghadam K, Kazemnejad S. Proliferation and chondrogenic differentiation potential of menstrual blood- and bone marrow derived stem cells in two-dimensional culture. Int J Hematol. 2012;95:484–93.

Meng X, Ichim TE, Zhong J, Rogers A, Yin Z, Jackson J, Wang H, Ge W, Bogin V, Chan KW, Thébaud B, Riordan NH. Endometrial regenerative cells: a novel stem cell population. J Transl Med. 2007;5:57.

Nikoo S, Ebtekar M, Jeddi-Tehrani M, Shervin A, Bozorgmehr M, Kazemnejad S, Zarnani AH. Effect of menstrual blood-derived stromal stem cells on proliferative capacity of peripheral blood mononuclear cells in allogeneic mixed lymphocyte reaction. J Obstet Gynaecol Res. 2012;38:804–9.

Patel AN, Park E, Kuzman M, Benetti F, Silva FJ, Allickson JG. Multipotent menstrual blood stromal stem cells: isolation, characterization and differentiation. Cell Transplant. 2008;17:303–11.

Zemel'ko VI, Grinchuk TM, Domnina AP, Artsybasheva IV, Zenin VV, Kirsanov AA, Bichevaia NK, Korsak VS, Nikol'skiĭ NN. Multipotent mesenchymal stem cells of desquamated endometrium: isolation, characterization and use as feeder layer for maintenance of human embryonic stem cell lines. Tsitologiia. 2011;53:919–29.

2.3.2 *Phenotypic Characterization of Endometrial Tissue-Derived Stem Cells*

Similar to the phenotypic characterization of menstrual blood-derived cells, the cells of endometrial tissue biopsy also possess positivity over wide range of markers; however, the marker percentage may vary according to its expression pattern, either could be remarkably expressed or moderately expressed or shows a low/sparse expression. As understood, the endometrial basalis located within the tissue is responsible for the regeneration of the functionalis layer that is shed during menstruation. Thus, with its dynamic characteristics and remodeling capacity of the endometrium, it becomes most imperative to characterize the marker profiles of endometrial tissue biopsy obtained through hysterectomies.

Several study analyzed the cell surface markers for the cultured endometrial stem cells derived from both functionalis and basalis layer of the endometrium for phenotypic expression. Hematopoietic stem cell markers including CD45, CD 14, CD19, CD56/16, CD34, and CD3 showed a negative expression, whereas markers which have been used to partially purify MSCs including CD90, CD105, CD73, STRO-1, CD146, CD9, CD13, CD29, CD44, CD117, CD133, Vimentin, and proMMP3 and so on stained positive, strongly suggesting the mesenchymal nature of the cells.

Besides, the other markers such as (CD34+CD45+) co-expressing CD7 and CD56 have been identified in human endometrial cell suspensions. Besides, some novel markers not reported regularly have also been identified by some researchers, for example, Musashi-1, an RNA-binding protein in neural stem cells and an epithelial progenitor cell marker. Expression of NAC1 in the glandular cells of various stages of endometrium was also noted. MSI1 and NOTCH1, which maintain stem cells in an undifferentiated state, tissue non-specific alkaline phosphatase, leucine-rich repeat containing G protein-coupled receptor-5 (Lgr-5), were also found to be present.

OCT-4, a highly expressed transcription factor in embryonic stem cells and in embryos at various stages of development, was also studied by researchers on endometrium. The expression of OCT-4 suggests the existence of endometrial stem cells. Its expression in both follicular phase and luteal phases of endometrium suggests the presence of pluripotent cells in the endometrium. The studies indicated that OCT-4 expression does vary with age, phase of the menstrual cycle, or with the gynecologic disorders of the individual. A thorough study on OCT-4 expression of the endometrial stem cells is needed and may help us understand the pathology of endometrial cancers and other disorders associated with abnormal proliferation of endometrium. Besides, stem cell markers, CD117 and CD34, were expressed throughout the menstrual cycle in human endometrial stroma.

Schwab and her co-workers demonstrated that EnSCs could be purified on the basis of their co-expression of two perivascular markers CD140b and CD146. Briefly, the stromal cell fraction obtained after enzymatic dissociation of endometrial tissue contained a mixture of cell populations that included CD45+ leukocytes and EPCAM+ epithelial cells. The stromal population was separated based on: MCAM (CD146)+/PDGFRB+ (eMSCs); MCAM (CD146)−/PDGFRB+ (endometrial stromal fibroblasts); and MCAM (CD146)+/PDGFRB− (endometrial endothelial cells). The MCAM (CD146)−/PDGFRB+ (stromal fibroblast) consisted of more endometrial stromal cells, whereas the MCAM (CD146)+/PDGFRB+ (eMSC) and MCAM (CD146)+/PDGFRB− (endothelial) populations comprised a small portion of the stromal fraction. Thus, eMSCs identified as above were also clonogenic with an MSC phenotype and were also able to differentiate into adipocytes, chondrocytes, and osteocytes and express MSC markers CD29, CD44, CD73, CD90, and CD105. Therefore, PDGF-Rβ+/CD146+ double positivity was suggested as a marker for endometrial mesenchymal stem cells (eMSCs). PDGFRβ+/CD146+ eMSCs exhibit pericyte characteristics and are localized adjacent to endometrial blood vessels.

Later, Thy-1 (CD 90) was used in conjugation with CD45 (negative marker) for clonogenicity. Along with CD90, CD146 was also found to enrich human endometrial CFU, suggesting the perivascular location of these cells. Thus, CD90 along with CD146 could be a potent marker for the sorting strategy of endometrial CFU. Recently, another marker, W5C5 (SUSD2) that is expressed in bone marrow MSCs, smooth muscle cells, and neoplastic cells, was introduced as an alternative single marker to purify the same PDGF-Rβ+/CD146+ eMSC population. It was also found to differentiate into adipogenic, osteogenic, chondrogenic, and myogenic cell lineages. Besides, W5C5+ eMSCs reconstituted endometrial stromal tissue after transplantation under the kidney capsule of non-obese diabetic/severe combined immunodeficient/IL-2 receptor γ-deficient (NOD/SCIDγ) mice. Furthermore, endometrial stem cells also found to express aSMA, but not CD31, indicating that cells are vascular smooth muscle cells and that they occupy a perivascular niche.

Although multitude of markers have been identified to isolate endometrial cells, the varied marker population between the heterogeneous endometrial cells and the culture purified eMSCs are poorly understood. Thus, our team carried out a study to identify the prospective markers of heterogeneous cell population isolated from the endometrial tissue, as well as the cultured endometrial cell population. Cell surface profile of heterogeneous non-cultured endometrial cells with respect to various markers including HSC (CD34, CD45, CD133, CD117), MSC (CD90, CD105, CD73), cell adhesion molecules (CD29, CD44, CD166, CD106, CD49d, CD31, CD54, CD13), markers of pluripotency (SSEA4, ABCG2, OCT4, SOX2), and perivascular markers (CD140b, CD146) were characterized using flowcytometry. Heterogeneous cell population of endometrial tissue was found to possess remarkable expression of CD29, higher expressions of CD44, CD73, CD140b, and moderate expressions of CD166, CD54, CD34, CD45, CD90, CD105, CD146, SSEA4, and OCT 4. However, the markers CD106, CD49d, CD31, CD13, CD133, and CD117 were expressed low, while SOX2 was identified to be sparsely expressed.

The characterization was also done upon cultures at both early (P1, P3, and P5) and late passages (P10, P15, P20, P25, and P30) of endometrial stem cells for various markers as specified above. This is the first of its kind study to characterize the markers even beyond passage 30. The marker expression profiles of endometrial stem cells upon culture are shown in Fig. 2.2. At early passages, cells were remarkably (>90 %) expressed by mesenchymal panel of markers, CD90, CD73, and CD105, and it seemed to be constant throughout the passages. Both CD105 and CD73 were remarkable from passage 1 and were consistently maintained throughout the remaining passages. CD90 was found to express high (75–89 %) throughout the early passages. Other non-stem cell markers were also assessed and found consistent with low (<25 %) expression for the markers, such as CD34, HLA-DR,

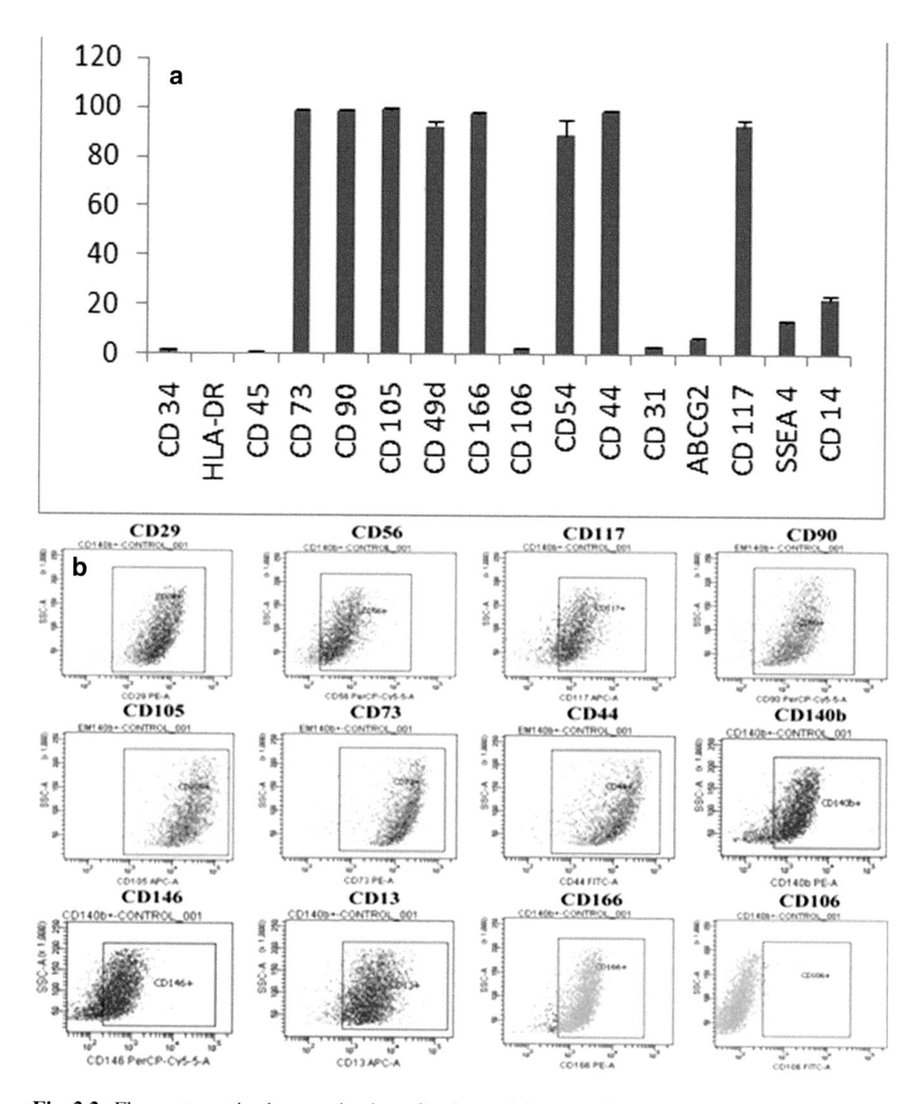

Fig. 2.2 Flowcytometric characterization of endometrial stem cells. Flowcytometric characterization of endometrial stem cells under culture condition. It shows remarkable expressions of various mesenchymal, perivascular, and pluripotent markers of endometrium. (**a**) Showing the mean ± SEM values of the expression profiles. (**b**) Data plots of flowcytometric analysis

CD117, and CD14. Though CD34 expression was found to be moderate initially, it showed a drastic reduction from the passage 2 onwards and subsequently reduced to low and same profile was consistently maintained throughout the early passages. Cell adhesion molecules being imperative for the functioning of cells were assessed for their variable expressions. Hematopoietic/endothelial migration-related markers such as CD31 as well as CD106 were found to show low expression, while the markers required for the general functioning of the cells revealed higher expression and was consistent throughout the early passages. These markers include CD54, CD49d, CD166, and CD29. Besides, the cells were also analyzed for their expression toward pluripotency or side population markers (SSEA4 and ABCG2) and was found to be consistently low throughout the early passage. Incidentally, the pericyte marker influenced a remarkable expression of CD140b. Nevertheless, the CD146 expression was found to be low throughout the early passages, indicating that the early passage cells are partially eMSCs, but most of them might be stromal fibroblasts, as reported that is also supported by Schwab and her co-workers.

Late passage analysis of endometrial cells also recorded a remarkable expression of mesenchymal stem cell markers (CD90, Cd105, and CD73) that was consistent with a lower expression of hematopoietic markers. Cell adhesion molecule expressions remained consistent as seen in the early passage. The expression of CD31 and CD106 was consistently sparse, and the high to remarkable expressing markers CD54, CD49d, CD166, and CD29 remained more or less constant with slight variations. The pluripotent/side population also found to be consistent with both SSEA4 and ABCG2 remained sparsely expressing. Though CD117 showed a lower expression in early passages, it showed an increase in expression toward the later passages, indicating the formation of pure eMSCs in the later passages. Besides, the CD146+ cells, which were low/sparsely present in early passage, became higher in the later passage, thereby further rationalizing the fact that the stromal fibroblast at early passage cultures are becoming pure eMSC population upon continuous culturing. This proves that there could a dedifferentiation approach taking place upon endometrial cell cultures. Dedifferentiation is a process by which cells develop in reverse, from a more differentiated to a less differentiated state. Understanding the process of dedifferentiation had bee a key element in today's research; in the future, definitive marker for dedifferentiation could evolve. This could be an ideal approach in studying stem cells, especially in a dynamic source such as endometrium.

Overall, the expressions exhibited in our study with regard to certain markers such as CD90, CD105, CD73, and other cell adhesion molecules were comparable with that of existing literature of Schwab and Gargett. Earlier studies of the perivascular cells of the endometrium have demonstrated its high colony-forming and multi-lineage differentiation capability. Similar to these reported studies, the present investigation also reported a higher expression of CD140b and CD146 markers in freshly isolated endometrial samples as well as cultured samples. This proves that the pericyte populations are eMSCs, thereby favoring the potent tissue repair mechanism in vivo. We also found a remarkable expression of nestin also in our study. The study demonstrated that human endometrium contains potentially pluripotent markers such as Oct4, SSEA4, and nanog. This explicate the fact that human endometrium may contain pluripotent cells. Our studies were comparable with that of certain literatures in relation to OCT4. Although the endometrium expresses as certain markers of pluripotency in the samples, it has to be acknowledged that there is no convincing evidence to conclude that the cells expressing these markers are pluripotent. However, it is well documented that pluripotent cells could grow well at late passages without losing its functionality. On a contrary, the multi-potent stem cells were shown to lose its property beyond certain passages in vitro. Thus, besides phenotypic characteristics, the study on pluripotent nature of this endometrial cell in view of its proliferation and differentiation potential at long-term culturing is also important. This was also demonstrated by our lab similar to other researchers and has been discussed in the forthcoming section.

Further Reading

Cervelló I, Gil-Sanchis C, Mas A, Delgado-Rosas F, Martínez- Conejero JA, Galán A, Martínez-Romero A, Martínez S, Navarro I, Ferro J, Horcajadas JA, Esteban FJ, O'Connor JE, Pellicer A, Simón C. Human endometrial side population cells exhibit genotypic, phenotypic and functional features of somatic stem cells. PLoS One. 2010;5:e10964.

Cho NH, Park YK, Kim YT, Yang H, Kim SK: Lifetime expression of stem cell markers in the uterine endometrium. Fertil and Steril 2004, 81(2):403–407.

Deepa Bhartiya, Ambreen Shaikh, Punam Nagvenkar, Sandhya Kasiviswanathan, Prasad Pethe, Harsha Pawani, Sujata Mohanty, S.G. Ananda Rao, Kusum Zaveri, and Indira Hinduja (2012). Very small embryonic-like stem cells with maximum regenerative potential get discarded during cord blood banking and bone marrow processing for autologous stem cell therapy. *Stem Cells Dev* 21, 1–6.

Dhanasekaran M, Indumathi S, Lissa RP, Harikrishnan R Rajkumar JS, and Sudarsanam D. (2012) A comprehensive study on optimization of proliferation and differentiation potency of bone marrow derived mesenchymal stem cells under prolonged culture condition. *Cytotechnology* Doi: 10.1007/s10616-012-9471-0.

Dimitrov R, Timeva T, Kyurkchiev D, Stamenova M, Shterev A, Kostova P, Zlatkov V, Kehayov I, Kyurkchiev S: Characterization of clonogenic stromal cells isolated from human endometrium. Reproduction 2008, 135(4):551–558.

Dominici M, Le Blanc K, Mueller I, Slaper-Cortenbach I, Marini F, Krause D, Deans R, Keating A, Prockop D, Horwitz E: Minimal criteria for defining multipotent mesenchymal stromal cells. The International Society for Cellular Therapy position statement. Cytotherapy 2006, 8(4):315–317.

Gargett CE, Chan RWS, Schwab KE (2007). Endometrial stem cells. Reproductive endocrinology 19(4), 377–383.

Ikoma T, Kyo S, Maida Y, Ozaki S, Takakura M, Nakao S, Inoue M. Bone marrow-derived cells from male donors can compose endometrial glands in female transplant recipients. Am J Obstet Gynecol. 2009;201(608):e1–8.

Indumathi S, Harikrishnan R, Rajkumar JS, Sudarsanam D and Dhanasekaran M. (2013) Prospective biomarkers of stem cells of human endometrium and fallopian tube in comparison to bone marrow. *Cell Tissue Res* 352(3), 537–549.

Lynch L, Golden-Mason L, Eogan M, O'Herlihy C, O'Farrelly C: Cells with haematopoietic stem cell phenotype in adult human endometrium: relevance to infertility? Human Reprod 2007, 22(4):919–926.

Masuda H, Anwar SS, Bühring H-J, Rao JR, Gargett CE: A novel marker of human endometrial mesenchymal stem-like cells. Cell Transplant 2012, 21(10):2201–2214.

Matthai C, Horvat R, Noe M, Nagele F, Radjabi A, van Trotsenburg M, Huber J, Kolbus A. Oct-4 expression in human endometrium. Mol Hum Reprod. 2006;12:7–10.
Mendez-Ferrer S, Michurina TV, Ferraro F, Mazloom AR, MacArthur BD, Lira SA, Scadden DT, Maayan A, Enikolopov GN, Frenette PS (2010). Mesenchymal and haematopoietic stem cells form a unique bone marrow niche. *Nature* 466, 829–834.
Miura M, Miura Y, Padilla-Nash H. M, Molinolo A. A, Fu B, Patel V, Seo B.-M, Sonoyama W, Zheng J. J, Baker C. C, Chen W, Ried T. & Shi S. (2006). Accumulated Chromosomal Instability in Murine Bone Marrow Mesenchymal Stem Cells Leads to Malignant Transformation. *stem cells* 24, 1095–1103.
Nguyen HP, Sprung CN, Gargett CE. Differential expression of Wnt signaling molecules between pre- and postmenopausal endometrial epithelial cells suggests a population of putative epithelial stem/progenitor cells reside in the basalis layer. Endocrinology. 2012;153:2870–83.
Schwab K, Gargett C (2007). Co-expression of two perivascular cell markers isolates mesenchymal stem-like cells from human endometrium. Hum Reprod, 22(11):2903–2911.
Schwab KE, Hutchinson P, Gargett CE (2008). Identification of surface markers for prospective isolation of human endometrial stromal colony forming cells. Human Reproduction 23, 934–943.
Schüring AN, Schulte N, Kelsch R, Röpke A, Kiesel L, Götte M: Characterization of endometrial mesenchymal stem-like cells obtained by endometrial biopsy during routine diagnostics. Fertil Steril 2011, 95(1):423–426.
Spitzer TLB, Rojas A, Zelenko Z, Aghajanova L, Erikson DW, Barragan F, Meyer M, Tamaresis JS, Hamilton AE, Irwin JC, Giudice LC (2011). Perivascular Human Endometrial Mesenchymal Stem Cells Express Pathways Relevant to Self-Renewal, Lineage Specification, and Functional Phenotype. *Biology of Reproduction* 86(2), 58.
Taylor HS. Endometrial cells derived from donor stem cells in bone marrow transplant recipients. JAMA. 2004;292:81–5.
Vikram Saapabathy, Saranya Ravi, Vivi Srivastava, Alok Srivastava, Sanjay Kumar (2012). Long-term cultured human term placenta-derived mesenchymal stem cells of maternal origin displays plasticity. doi:10.1155/2012/174328.

2.4 Proliferation and Multi-potent Differentiation

The Mesenchymal and Tissue Stem Cell Committee of the International Society for Cellular Therapy (ISCT) has put forward certain minimal criteria to define and identify the MSC population. One of the criteria to define MSCs is their ability to undergo self-renewal and multi-lineage differentiation. In brief, the cells must be able to differentiate to mesodermal lineages of osteoblasts and adipocytes to demonstrate bone and fat phenotypes, respectively, under standard in vitro differentiating conditions. Evidence on the existence of endometrial stem cells was derived from the phenotypic, functional, and proliferative studies. Emerging evidence suggests that human and mouse endometrium contains adult stem cells. As per the criteria put forth by ISCT, several studies have put effort to understand whether the stem cells of endometrium have the ability to proliferate and differentiate into multiple lineages as described for other stem cell types.

Differentiation is the process which leads to the expression of phenotypic properties characteristic of the functionally mature cell in vivo. Ideal conditions inducing the process of differentiation include a high cell density, enhanced cell-cell and cell-matrix interactions, and the presence of various differentiation factors. The biological property that most uniquely identifies MSCs is their ability to differentiate into a variety of connective tissue cell types, such as osteocytes, adipocytes, and chondrocytes. This trilineage differentiation capacity of MSC makes it an ideal candidate for clinical tissue regeneration strategies. The commitment and differentiation of MSC to specific mature cell type is influenced by the activities of various transcription factors, cytokines, growth factors, and extra cellular matrix (ECM) molecules. Under defined inductive condition, MSCs are capable of acquiring characteristic of cells derived from embryonic mesoderm such as osteoblast, adipocytes, and chondrocytes. Differentiation to osteoblasts can be demonstrated by staining with Alizarin Red or von Kossa staining. Adipocyte differentiation is most readily demonstrated by staining with Oil Red O. Chondroblast differentiation is demonstrated by staining with Alcian blue or immunohistochemical staining for collagen type II.

The potency of mesenchymal stem cells was considered being capable of only forming tissues of mesodermal origin, as has been demonstrated. However, in recent years mesenchymal stem cells, originated in the mesoderm, were identified to cross the germ layer boundary and undergo different mesenchymal lineage. This process has been termed as transdifferentiation or plasticity. This concept that mesenchymal stem cells possess a far wider potential of transdifferentiation is also supported from the various reported literatures. This perception of the far wider potential of mesenchymal stem cells has led to considerable excitement with regard to its potential therapeutic applications in regenerative medicine and tissue engineering. Similar to other sources of stem cells, the endometrial stem cells, from both menstrual blood- and endometrium-derived stem cells, have created a great excitement. As per the criteria put forth by ISCT, it is imperative to understand whether the multi-potent differentiation ability of EnScs as that of other stem cell types. Many researchers have shown both these stem cells have multi-potent differentiation ability, as described below.

Further Reading

Antono Uccelli, Lorenzo Moretta, Vito Pistoia (2008). Mesenchymal stem cells in health and disease. *Nature reviews immunology* 8, 726–736.

Dominici M, Le Blanc K, Mueller I, Slaper-Cortenbach I, Marini F, Krause D, Deans R, Keating A, Prockop Dj, Horwitz E (2006). Minimal criteria for defining

multipotent mesenchymal stromal cells The International Society for Cellular Therapy position statement. *Cytotherapy* 8(4), 315–317
Pittenger MF, Mackay AM, Beck SC, Jaiswal RK, Douglas R, Mosca JD, Moorman MA, Simonetti DW, Craig S, Marshak DR (1999) Multilineage Potential of Adult Human Mesenchymal Stem Cells. *Science 284*(5411), 143–147

2.4.1 Multi-potent Differentiation of Menstrual Blood-Derived Stem Cells

Several independent studies have clearly demonstrated the multi-potent differentiation ability of MenSCs. Meng et al. maintained the "endometrial regenerative cells" for more than 68 doublings without showing karyotypic abnormalities. The multipotency of MenSCs has been demonstrated by directly differentiating them into chondrogenic, adipogenic, osteogenic, neurogenic, and cardiogenic cell lineages using the specific human mesenchymal stem cell differentiation bullet kit. Tissue-specific markers of target cells were detected at the cellular and molecular levels. These studies laid the status of application of MenSCs in regenerative medicine. Similar work done by Meng et al. has demonstrated that MenSCs are capable of differentiating in standard culture reagents into nine lineages: cardiomyocytic, respiratory epithelial, neurocytic, myocytic, endothelial, pancreatic, hepatic, adipocytic, and osteogenic. They found that these stem cells can produce matrix metalloproteinase-3 (MMP3), MMP10, granulocyte macrophage colony-stimulating factor (GM-CSF), angiopoietin-2, and platelet-derived growth factor (PDGF)-BB in quantities 10–10,000 times higher than umbilical cord blood cells.

Many other researchers have worked on the differentiation potential of eMSCs into various lineages of ectodermal, endodermal, and mesodermal lineages. Besides, the eMSCs have been shown to differentiate into endothelial lineages. Some researchers have also shown that MenSCs are able to differentiate into cardiocytes with the functions of beating spontaneously after induction and decreasing the myocardial infarction area in a rat model. After a one-month proliferation, mesenchymal cells from nine different subjects' menstrual blood co-cultured for 3 days with rat cardiocytes, began to beat naturally, exhibited cardiomyocyte-specific action potentials, and eventually transformed into a sheet of cardiocytes. The nurtured cells were transplanted into mice with myocardial infarction, which significantly improved their condition. MenSCs transferred dystrophin into dystrophied myocytes through cell fusion and transdifferentiation in vitro and in vivo. Vascular injection of cells into the sites of ischemia significantly improved limb ischemia with angiogenesis in the ischemic area. This has also been confirmed by Murphy et al. (2008) using MenSCs. The MenSCs in mice ischemic segments displayed the following characteristics: production of high level of growth factor and matrix metalloproteinase, inhibition of the inflammatory response, and proliferation and differentiation without losing any function. This illustrates not only its in vitro differentiation ability,

Germ Layers	Induction Medium for MeSC	Gene highly expressed
	Mesodermal lineage	
Adipocyte	(D)MEM 10% FBS, 1 umol/L dexamethasone, 60 umol/L indomethacin and 50 umol/L 3-isobutyl-1-methyl-xanthine IBMX **(Alcayaga-Miranda et al., 2015).** 10 μg/ml recombinant human insulin, DMEM-LG/GL, 1% penicillin/streptomycin **(Patel et al., 2008).**	peroxysome proliferator-activated receptor (PPAR)-γ
Chondrocyte	DMEM, 0.1 uM dexamethasone, 0.17 mM ascorbic acid, 1% insulin–transferring selenic acid (ITS) and 10 ng/ml transforming growth factor b3 (TGFb3) **(Alcayaga-Miranda et al., 2015).**	aggrecan and collagen II
Osteocyte	DMEM 10% FBS, 0.1 umol/L dexamethasone and 50 ug/mL ascorbic acid in the presence or absence of 3 mM of Na2HPO4) **(Alcayaga-Miranda et al., 2015).**10 mM β-glycerophosphate, 0.2 mM ascorbate, and 5% FBS **(Patel et al.,2008).**	osteocalcin (OC)
Cardiogenic/ Myogenic	LG-DMEM, 20% FBS collagen I–coated dishes, 5 or 10 μM 5-azacytidine, 2% horse serum (HS) or 1% insulin-transferrin-selenium supplement [ITS] **(Cui CH et al., 2007; Toyoda M et al., 2007).**1% penicillin/streptomycin, 2mM Glutamax, 8 μM 5-aza-2'-deoxycytidine (Aza) or 400–800 μM S-nitroso-N-acetylpenicillamine (SNAP) **(Patel et al., 2008).**	MYoD, desmin, myogenin and Connexin 43

Fig. 2.3 Mesodermal differentiation lineage protocol of menstrual blood stem cells. Induction medium of mesodermal differentiation of menstrual blood stem cells by various authors are summarized

but also its in vivo differentiation ability. The detail differentiation protocol carried out by various researchers is summarized (Figs. 2.3 and 2.4).

Further Reading

Alcayaga-Miranda, Jimena Cuenca, Patricia Luz-Crawford et al. Characterization of menstrual stem cells: angiogenic effect, migration and hematopoietic stem cell support in comparison with bone marrow mesenchymal stem cells. Stem Cell Research & Therapy, 2015; 6:32.

Cui CH, Taro Uyama, Kenji Miyado, et al. Menstrual Blood-derived Cells Confer Human Dystrophin Expression in the Murine Model of Duchenne Muscular Dystrophy via Cell Fusion and Myogenic Transdifferentiation. Molecular Biology of the Cell, 2007; Vol. 18, 1586–1594.

Hida N, Nishiyama N, Miyoshi S, Kira S, Segawa K, Uyama T, Mori T, Miyado K, Ikegami Y, Cui C, Kiyono T, Kyo S, Shimizu T, Okano T, Sakamoto M, Ogawa S, Umezawa A. Novel cardiac precursor- like cells from human menstrual blood-derived mesenchymal cells. Stem Cells. 2008;26:1695–704.

Meng X, Thomas E Ichim, Jie Zhong, et al. Endometrial regenerative cells: A novel stem cell population. Journal of Translational Medicine, 2007;5:57

Murphy P, Hao Wang, Amit N Patel, et al. Allogeneic endometrial regenerative cells: An "Off the shelf solution" for critical limb ischemia?. Journal of Translational Medicine, 2008; 6:45.
Patel, Eulsoon Park, Michael Kuzman et al. Multipotent Menstrual Blood Stromal Stem Cells: Isolation, Characterization, and Differentiation. Cell Transplantation, 2008; Vol. 17, pp. 303–311.
Toyoda M, CH. Cui, et al. Myogenic transdifferentiation of menstrual blood-derived cells. Acta Myologica, 2007; XXVI; p. 176–178
X Mou, Jian LIN, Jin-yang CHEN, et al. Menstrual blood-derived mesenchymal stem cells differentiate into functional hepatocyte-like cells. J Zhejiang Univ-Sci B (Biomed & Biotechnol) 2013; 14(11):961–972

2.4.2 Multi-potent Differentiation of Endometrial Tissue-Derived Stem Cells

The ability of CD146+PDGFRb+MSC like cells or clonogenic human endometrial stromal cells to differentiate into mesodermal origin such as adipocytes, osteocytes, smooth muscle cells and chondrocytes, endothelial cells has well been demonstrated.

Germ Layers	Induction Medium for MeSC	Gene highly expressed
	Ectodermal lineage	
Cholinergic Neuron /Motor Neuron	DMEM-F/12, 1% penicillin/streptomycin, 2 mM Glutamax, 1 x N-2 suppplement, 10 ng/ml bFGF, 10 ng/ml platelet-derived growth factor (PDGF), 20 ng/ml epidermal growth factor (EGF) **(Patel et al., 2008).**	Nestin, NCAM, and Nurr-1.
	Endodermal lineage	
Hepatic-like cells	Iscove's Modified Dulbecco's Medium (IMDM), 0 ng/ml hepatocyte growth factor (HGF), 10 ng/ml fibroblast growth factor-4 (FGF-4), 10 ng/ml oncostain M, 40 μg/L dexamethasone, and 1 x ITS+ Premix **(X MOU et al., 2013)**. hFGF-4 (20 ng/ml), SCF (40 ng/ml) **(Meng X et al., 2007).**	albumin (ALB), α-fetoprotein (AFP), cytokeratin 18/19 (CK18/19), and cytochrome P450 1A1/3A4 (CYP1A1/3A4).
	Other lineage	
Angiogenesis	DMEM 2% FBS, exposed to normoxic and hypoxic conditions (1% O2), Conditional Media or in endothelial growth medium, 200 μl growth factor **(Alcayaga-Miranda et al., 2015; Murphy P et al., 2008).**	VEGF and bFGF

Fig. 2.4 Ectodermal and endodermal lineage differentiation protocol of menstrual blood stem cells. Induction medium of Ectodermal and endodermal lineage differentiation of menstrual blood stem cells by various authors are summarized

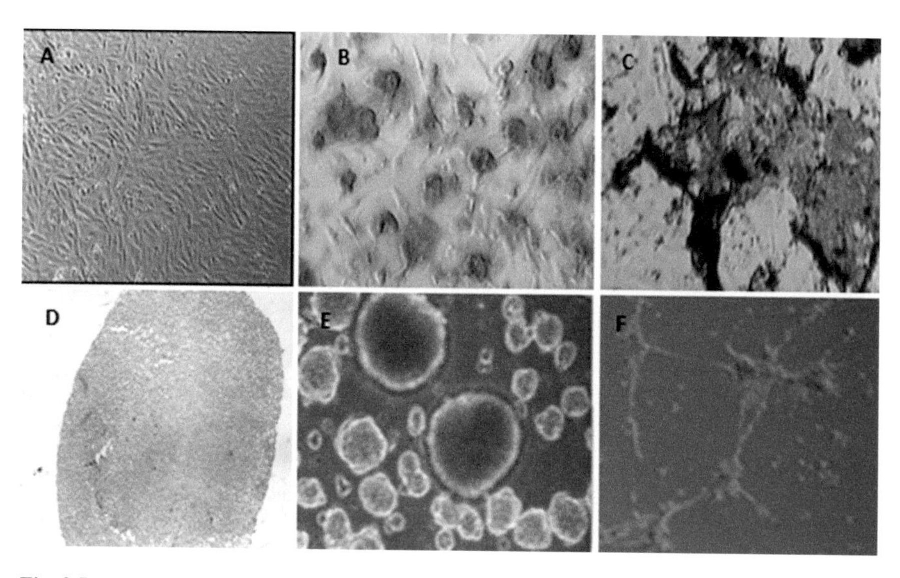

Fig. 2.5 Multi-differentiation potential of endometrial stem cells. (**a**) Confluent endometrial stem cells at culture, (**b**) osteoblasts (alizarin staining), (**c**) osteoblasts (von Kossa), (**d**) chondrogenic differentiation (*crystal violet*), (**e**) islet differentiation, (**f**) endothelial cells

It has been demonstrated that not only the endometrial stem cells but also the SP cells differentiate into endothelial and smooth muscle cells. The possibility of these endometrial stem cells to differentiate into neuronal like cells has been evaluated. Studies have extended also on differentiation of eMSCs into oligodendrocytes. Our laboratory investigated the differentiation potential of endometrial stem cells at P30 or beyond. We provide evidence that endometrial stem cells possess the ability to differentiate into osteocytes, adipocytes, chondrocytes, endothelial cells, and islet like clusters. We demonstrate that the endometrial stem cells could retain its ability to differentiate into all mesodermal lineages; however, we could not get the transdifferentiation ability of endometrial stem cells at later passage beyond P20. The differentiation potency of endometrial stem cells carried in our laboratory is illustrated (Fig. 2.5).

In a recent well-designed study, endometrial mesenchymal stem cells (eMSCs) were induced to produce not only mesodermal and neuroectodermal lineage cells, but also pancreatic lineage cells in the serum-free modified pancreatic selection media. Microarray analysis showed that the expression levels of 716 genes changed between the primary and induced cells. Furthermore, a streptozotocin (STZ)-induced animal model of diabetes was used to demonstrate the ability of spheroid-like body (SB) EMSCs to engraft, differentiate in vivo, regulate blood glucose levels, and significantly prolong the survival of graft cells. Hugh Taylor's group had also shown that endometrial mesenchymal stem cells could be success-

Germ Layers	Induction Medium for EMSC	Gene highly expressed
Mesodermal lineage		
Adipocyte	Iscove's modified Dulbecco's medium, 0.5 mM 3-isobuty1-1-methylxanthine, 1 μM hydrocortisone , 0.1 mM indomethacin , and 10% rabbit serum **(Yang Li et al., 2010).** Isobuty1-1-methylxanthine 500μM, Dexamethasone 1μM, Insulin 10μM, indomethacin 200μM **(Gargett et al., 2009; Jafar Ai et al., 2010).**	PPAR gamma2 mRNA , Lipoprotein lipase (LPL), PPARα
Chondrocyte	HG-DMEM, 0.1 μM dexamethasone, 50μg/ml ascorbic acid, 100μg/ml sodium pyruyat , 40μg/ml proline , 10 ng/ml transforming growth factor-ß1, and 50 mg/ml ITS+ premix (6.25 μg/ml insulin, 6.25 μg/ml transfemin, 6.25 ng/ml selenious acid, 1.25 mg/ml bovine serum albumin, and 5.35 mg/ml linoleic acid) **(Yang Li et al., 2010; Wolff et al., 2007).** Insulin 6.25μg/ml, Ascorbic Acid 2- phosphate 50μM, transforming growth factor-ß1 10ng/ml. **(Gargett et al., 2009).**	Col2a1 (a chondrogenic marker), Collagen Type II
Osteocyte	Osteogenic medium consits of DMEM-LG supplemented with 50 lg/ml ascorbate-2 phosphate, 10^{-8} M dexamethasone, and 10 mM ß-glycerophosphate. **(Yang Li et al., 2010; Chen et al., 2010).** 1α-25-dihydroxyvitamin-D3 0.01μM, Ascorbic Acid 2-phosphate 50μM, ß-glycerophosphate 10mM **(Gargett et al., 2009).**	Runx2(a osteogenic marker), Parathyroid hormone receptor 1 (PTHR 1)
Smooth muscle	M199 (DMEM)/F12 medium, 2 mM L-glutamine, 50 U/mL penicillin, 50 mg/ml streptomycin, 5 ng/mL TGF-b1, 3 ng/mL platelet-derived growth factor (PD GF-BB), 3 ng/mL hepatocyte growth factors (HGF) and 5 ng/mL vascular endothelial growth factor (VEGF) **(Shoae-Hassani et al. 2013).**	vinculin, a-SMA and calponin
Cardiogenic/Myogenic	DMEM-F12, 2% FBS, 100 U/mL penicillin, 100 μg/ml streptomycin , and 10 μM 5-azacytidine**(Faezeh et al., 2014).** Myogenic (MM) DMEM FBS (10%), Heat inactivated male Human Serum (5%), 50 M hydrocortisone, 1% antibiotic/antimycotic **(Gargett et al.,2009).**	Desmin, MyoD, Troponin, and caldesmon

Fig. 2.6 Mesodermal differentiation lineage protocol of endometrial stem cells. Induction medium of mesodermal differentiation of endometrial stem cells by various authors are summarized

fully reprogrammed into insulin-producing cells in vitro. The study was confirmed by various molecular and histochemical analyses. Therapeutic purposes of these cells in patients with type 1 diabetes were apparent by its effectiveness in a STZ-induced model of diabetes. Mice transplanted with differentiated cells displayed no gross pathological symptoms and showed a stabilization of glucose. In contrast, non-transplanted control diabetic mice or mice transplanted with undifferentiated hEMSCs showed a large number of complications in the initial observation period. These studies have unearthed the prospects of curative therapeutics. The differentiation of eMSCs into different lineages had been demonstrated (Figs. 2.6, 2.7 and 2.8).

Germ Layers	Induction Medium for EMSC	Gene highly expressed
	Ectodermal lineage	
Cholinergic Neuron /Motor Neuron	DMEM/F12 (1:1) Serum free media and 0.6% glucose, 25 μg/ml insulin, 100μg/ml ransferring, 20 nM progesterone, 60 μM putrescine, 30 nM selenium chloride, 2 mM glutamine, 3 mM sodium bicarbonate, 5mM HEPES, 2 μg/ml heparin, 20 ng/ml EGF,and 20 ng/ml bFGF, 20 ng/ml Sonic hedgehog , 10 ng/ml brain-derived neurotrophic factor, and all-trans retinoic acid 100 nM **(H-Yang Li et al., 2010: Noureddini M et al,2012)** .0.2 %B27, 100 ng/ml of Shh, and 0.01 ng/ml RA, 100 ng/ml glial cell-derived neurotrophic factor (GDNF), and 200 ng/ml brain-derived neurotrophic factor (BDNF) **(S. Ebrahimi-Barough et al.,2014).**	ChAT(choline acetyltransferace MAP2 (microtubule associated protein 2),NF-1 (neurofilament L), Islet-1,and HB9.
Oligodendrocytes	serum-free N2 medium DMEM/F12, 1 %N2 , 2 mML-glutamine, and penicillin/streptomycin, 20 ng/mL basic fibroblast growth factor (FGF) 2, 20 ng/mL epidermal growth factor, 20 ng/mL platelet-derived growth factor AA,30 ng/mL triidothyronine and 40 % neuronal condition media **(S.Ebrahimi-Barough et al.,2013)**	Nestin and PDGFRα

Fig. 2.7 Ectodermal lineage differentiation protocol of endometrial stem cells. Induction medium of ectodermal differentiation of endometrial stem cells by various authors are summarized

Further Reading

Esfandiari Navid, D.V.M., Mozafar Khazaei, Jafar Ai, et al. Angiogenesis Following Three-Dimensional Culture of Isolated Human Endometrial Stromal Cells. International Journal of Fertility and Sterility, 2008; Vol 2, No 1, Pages: 19–22.

Faezeh Faghihi, Hora Jalali, Kazem Parivar, et al. Evaluation of Differentiation Potential of Endometrial- Versus Bone Marrow- Derived Mesenchymal Stem Cells Into Myoblast-Like Cells. International Journal of Current Life Sciences, 2014; Vol.4, Issue, 8, pp. 3992–3997.

Gargett Caroline E., Kjiana E. Schwab, Rachel M. Zillwood, et al. Isolation and Culture of Epithelial Progenitors and Mesenchymal Stem Cells from Human Endometrium. Biology Of Reproduction, 2009; 80, 1136–1145.

H-Yang Li, Yi-Jen Chen, Shih-Jen Chen, et al. Induction of Insulin-Producing Cells Derived from Endometrial Mesenchymal Stem-like Cells. JPET, 2010; 335:817–829.

Jafar Ai, A R Shahverdi, Somayeh Ebrahimi Barough, et al. Derivation of Adipocytes from Human Endometrial Stem Cells (EnSCs). J Reprod Infertil. 2012; 13(3):151–157.

Germ Layers	Induction Medium for EMSC	Gene highly expressed
	Endodermal lineages	
Pancreatic cells	DMEM(25 mMglucose)/F12 (1:1) Serum free medium , 0.6% glucose , 25 μg/ml insulin, 100 μg/ml transferrin, 20 nM progesterone, 60 μM putrescine, 30 nM selenium chloride, 2 mM glutamine, 3 mM sodium bicarbonate, 5 mM HEPES buffer, 2 μg/ml heparin, 20 ng/ml EGF, 20 ng/ml bFGF, and 20 ng/ml hepatocyte growth factor, 10 nM nicotinamide, and 100 ng/ml Activin A **(H-Yang Li et al., 2010).** 10% FBS and 10^{-6} mol/l retinoic acid, HG-DMEM, LG-DMEM 5.56 mmol/l, 300 nmol/l of indolactam V. 10 nmol/l exendin-4, 1% penicillin/streptomycin and 1% amphotericin B **(Xavier S et al., 2011).**	insulin, Glut2, Pax4,Nkx2.2, NeuroD,Isl-1, somatostatin, PDX1 and glucagan
Hepatic-like cells	hepatocyte growth factor (40 ng/ml), b-FGF (20 ng/ml), hFGF-4 (20 ng/ml), SCF (40 ng/ml) **(F. Khademi et al. 2014)** oncostatin M,and trichostatin A **(YangXY et al., 2014)**	albumin and cvtokeratin 8
Germ Layers	**Induction Medium for EMSC**	**Gene highly expressed**
	Other Lineages	
Macrophage	Iscove's modified dulbecco's medium, 50 ng/ml TPO, 0.5% BSA , 200 μg/ml iron saturated transferring, 10 μg/ml insulin, 2 mM L-glutamine, 4 μg/ml LDL cholesterol, 50 μM 2-B-mercaptoethanol, 20 μM each nucleotide, 20 μM dNTP, and 1% antibiotic-antimyotic solution **(Wang J et al., 2012).**	P-selectin (CD62p)
Angiogenesis	1 ml/well fibrinogen solution (3mg/ml in M199 culture medium) and 15 ul thrombin (gel formation), 1ml M199 supplemented with 5% fetal bovine serum (FBS), 0.1% e-amino-glutamine and 100 X antibiotic solutions. **(Esfandiari N et al., 2008; masuda H et al., 2007)**	CD31

Fig. 2.8 Endodermal and other lineage differentiation protocol of endometrial stem cells. Induction medium of endodermal and other differentiation of endometrial stem cells by various authors are summarized

K Kato, M Yoshimoto, K Kato et al. (2007). Characterization of side population cells in human normal endometrium. *Human Reproduction* 22, 1214–1223.

Khademi F, Javad Verdi, Masoud Soleimani, et al. Human endometrial adult stem cells can be differentiated into hepatocyte cells. Journal of Medical Hypotheses and Ideas, 2014; 8,30–33

Masuda H, Matsuzaki H, Hiratsu E et al., (2010). Stem cell- Like properties of the Endometrial side population: Implication in Endometrial regeneration. *Plos one* 5(4).

Masuda H, Tetsuo Maruyama, Emi Hiratsu, et al. Noninvasive and real-time assessment of reconstructed functional human endometrium in NOD/SCID/c null immunodeficient mice. PNAS, 2007; vol. 104 no. 6 1925–1930.

Noureddini M, Verdi J, Mortazavi Tabatabaei SA, et al. Human endometrial stem cell neurogenesis in response to NGF and bFGF. Cell Biol Int, 2012; 36: 961–6

S-Ebrahimi-Barough, Homa Mohseni Kouchesfahani, et al. Differentiation of Human Endometrial Stromal Cells into Oligodendrocyte Progenitor Cells (OPCs). J Mol Neurosci,2013.

S-Ebrahimi-Barough, Abbas Norouzi Javidan, et al. Evaluation of Motor Neuron-Like Cell Differentiation of hEnSCs on Biodegradable PLGA Nanofiber Scaffolds. Mol Neurobiol,2014.

Shoae-Hassani A, Shiva Sharif, Alexander M. Seifalian, et al. Endometrial stem cell differentiation into smooth muscle cell: a novel approach for bladder tissue engineering in women. BJU Int, 2013; 112: 854–863.

Tsuji S, Yoshimoto M, Kato K et al., (2008). Side population cells contribute to the genesis of human endometrium. *Fertil Steril* 90, 1528–1537.

Wang J, Chen S, Zhang C, Stegeman S, Pfaff-Amesse T, et al. Human Endometrial Stromal Stem Cells Differentiate into Megakaryocytes with the Ability to Produce Functional Platelets. PLoS ONE, 2012; 7(8): e44300.

Wolff E.F., A.B. Wolff, H. Du, et al. Demonstration of multipotent stem cells in the adult human endometrium by in vitro chondrogenesis. Report Sci, (2007); 14, 524–533.

Xavier S, Efi E Massasa, Yuzhe Feng, et al. Derivation of Insulin Producing Cells From Human Endometrial Stromal Stem Cells and Use in the Treatment of Murine Diabetes. Molecular Therapy, 2011; vol. 19 no. 11, 2065–2071.

Yang XY, Wang W, et al. In vitro hepatic differentiation of human endometrial stromal stem cells. In Vitro Cell Dev Biol Anim. 2014; 50(2):162–70.

Chapter 3
Applications of Endometrial Stem Cells

Treatments for various degenerative disorders rely on several treatment modalities like drugs through allopathy, homeopathy and ayurveda and surgical interventions that modify the system (Fig. 3.1). There are limitations and hindrance in these therapeutic approaches when it comes to regeneration of damaged tissues and cells. To overcome these obstacles, there emerged cell-based therapies, especially stem cells (Fig. 3.1). Regenerative medicine is a multidisciplinary field of research that is rapidly expanding with the technological advancements in recent years. It involves the use of stem cells, sometimes with the combination of stem cells and tissue engineering using growth factors and biomaterials that helps to cure the diseases by performing several functions of reparative, replacive, regenerative, or rejuvenative techniques to restore damaged, aging, or diseased cells, tissues, and organs. Detailed study on adult stem cells in the past decade has kindled useful knowledge about developmental, morphological, and physiological processes that form the basis of tissue and organ formation, maintenance, repair, and regeneration. Accordingly, general clinical, scientific, and public attention to the application of stem cell therapy has been substantial (Fig. 3.1). The correlation of stem cell technology with tissue repair still has a long way to go. Conceptually and from a practical standpoint, bone marrow has been the most influential source of stem cells that offers a possibility of being used in a wide range of therapeutics. This is due to the first successful transplantation of bone marrow-derived hematopoietic stem cells dates back to the late 1960s. It is known to possess heterogeneous stem cells that participate in all steps of tissue regeneration. Despite its colossal potential and benefits in certain diseases, stem cells derived from bone marrow have not been promising to attempt curative therapeutics for all diseases. Clinical situations frequently demand stem cells with dependable quality and quantity to treat disorders of cellular degeneration.

Challenges to bring stem cells from bench to bedside have expanded rapidly. Recent progress in stem cell biology has allowed researchers to investigate distinct stem cell populations in tissues and organs such as tendon, periodontal ligament, synovial membrane, lung, liver, endometrial tissue and body/tissue fluids such as

I. Somasundaram, *Endometrial Stem Cells and Its Potential Applications*, SpringerBriefs in Stem Cells, DOI 10.1007/978-81-322-2746-5_3

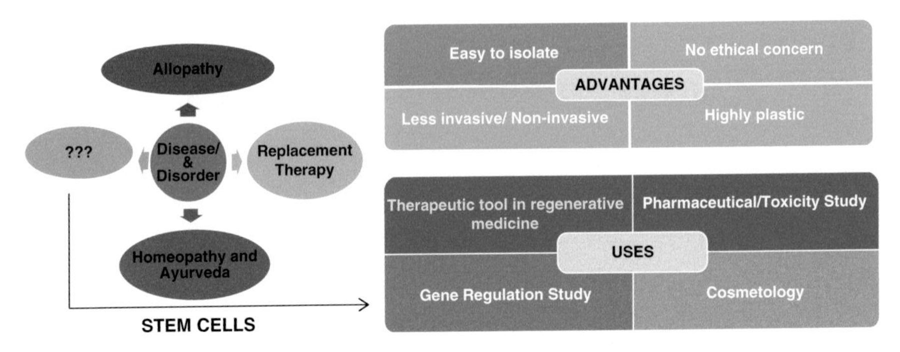

Fig. 3.1 Stem cells in disease and disorder. It represents that stem cells are the alternate safe therapy as compared to other existing in treating many diseases. Also, it highlights its application

synovial fluid, amniotic fluid, and menstrual blood. Although various sources of stem cells exist, endometrium is a dynamic source of stem cells, because a constant renewable source of stem cells could be obtained even post-menopausal conditions.

3.1 Applications in Regenerative Medicine: An Overview

The presence of stem cells in the endometrium was described about 30 years ago, from the observation that the upper layers of this tissue shed and were renovated each month, but the cells that were shed represented non-viable cells. In particular, the cells in the basalis were viable but not the shed cells from the functionalis. Only in the last few years, however, endometrial cells been well been characterized. Accumulating evidence has shown that the uterine endometrium contains multipotent adult stem cells that have the ability to proliferate in culture, express a consistent set of markers, and can be reproducibly differentiated to multiple non-endometrial cell lineages under controlled in vitro conditions.

Although bone marrow lasts till today as the most ideal source of stem cell transplantation in regenerative medicine, due to its disadvantages, such as invasive procedure, age-related decline in its potency, anesthesia and so on limits the applicability. Endometrial stromal cells possess wide range of advantages as opposed to other post-natal stem cells, including bone marrow to prove them as a valuable tool in cell-based therapies. They are as follows: easy to isolate, readily available, non-invasive, high accessibility, trash source, longer preservation, highly clonogenic with a higher multi-differentiation ability, possess immunomodulatory properties and inherent angiogenic potential. Besides, it could be easily restored with estrogen treatment after menopause. Additionally there are several hysterectomies performed yearly all over the world that could serve as a donor bank providing stem cells for men and for women. Besides, the endometrium provides a source of immunologically matched cells for tissue engineering without concern for rejection. Allickson et al. showed that infusion of MenSCs into Harlan Sprague Dawley mice and Dunkin Hartley albino guinea pigs for 7 days showed an absence of extraneous toxic contaminants, and all animals remained healthy, without reactions, and showed no weight loss.

These authors also showed that MenSCs demonstrate a moderate response in a mixed lymphocyte reaction. Besides, a clinical trial and an in vitro immunologic test demonstrated that MBPCs possessed low-immunogenicity properties and immunomodulatory effects. Thereby, endometrial stem cells serve as an ideal autologous/allogenic therapeutic tool in regenerative medicine. This is evidenced by several pre-clinical and clinical trials on endometrial stem cells in autologous/allogenic transplantations. Endometrial stem cells have already entered pre-clinical and clinical trial in treating various disorders such as peripheral arteries and vascular disorders, Parkinson's disease, diabetes, and stroke. Animal experiments showed that menstrual blood-derived stem cells had tissue repair effects in some diseases such as Duchenne muscular dystrophy (DMD), myocardial infarction (MI), critical limb ischemia (CLI), and stroke.

The first report of clinical use of ERC involved treating patients with multiple sclerosis. No adverse events were reported at the time of last follow-up. Similar lines of clinical trial were also demonstrated with muscular dystrophies and heart failures. An experimental study demonstrated that menstrual blood-derived stromal cells promoted functional improvement of damaged heart tissue, with evidence of cell engraftment and transdifferentiation into cardiac tissue. Borlongan et al. published the results of menstrual blood cell transplantation in experimental stroke. He demonstrated embryonic-like stem cell phenotypic markers, such as Oct4, SSEA, and Nanog, and other neural markers upon its culture. Wolff et al. reported the use of endometrial-derived neural cells in a Parkinson's disease immunocompetent mouse model. Migration, differentiation, and production of dopamine were detected in vivo, demonstrating its potential to restore the damaged tissue. Following this, several other studies are underway, and endometrial stem cells have a proven record of being a valuable ideal source of stem cell therapeutics. Some of its in vivo pre-clinical and clinical applications are discussed below in detail.

Apart from menstrual blood-derived cells, tissue-derived endometrial stem cells obtained by low invasive biopsy or hysterectomy have been used for therapeutic purposes in several animal models of disease/in clinical trials. For example, Hida et al. demonstrated the therapeutic efficacy of both MenScs and EnScs in vivo and in vitro. Wolff et al. also demonstrated the therapeutic efficacy of endometrial tissue-derived MSCs from nine reproductive-aged women. Using an in vitro differentiation protocol in the presence of FGF and EGF and other differentiation medium, the authors were able to obtain dopaminergic neurons. These cells were transplanted into mouse models of Parkinson's disease. They partially restored dopamine levels in vivo. Besides, the study conducted by Hugh Taylor demonstrated the potential of endometrial stem cells for therapeutic purpose in patients with type I diabetes. The experimental group of animals treated with human endometrial MSCs showed a stabilization of glucose. Thus, it is evident that endometrial stem cells possess ability to treat human diseases. Results of recent endometrial

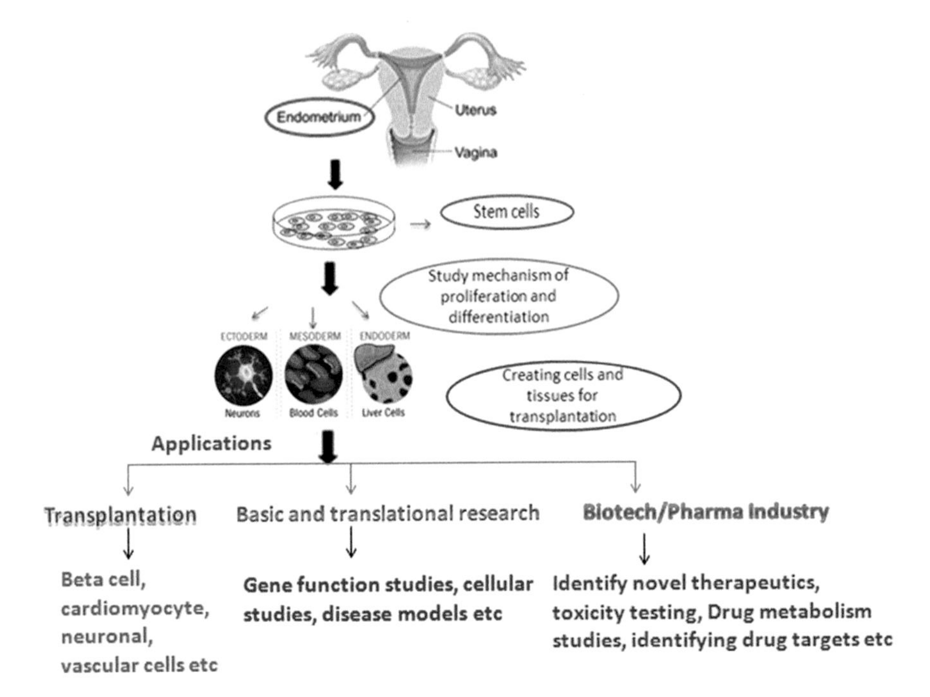

Fig. 3.2 Applications of endometrial stem cells: a nutshell. It represents in vitro and in vivo attributes of endometrial stem cells for its utility in various applications

stem cell research have received much clinical, scientific, and public attention, and clinical applications of endometrial stem cell therapy will undoubtedly continue to expand in the future (Fig. 3.2).

Further Reading

Allickson JG, Sanchez A, Yefi menko N, Borlongan CV, Sanberg PR. Recent studies assessing the proliferative capability of a novel adult stem cell identified in menstrual blood. Open Stem Cell J. 2011;3:4–10.

Blondheim NR, Levy YS, Ben-Zur T, Burshtein A, Cherlow T, Kan I, Barzilai R, Bahat Stromza M, Barhum Y, Bulvik S, Melamed E, Offen D. Human mesenchymal stem cells express neural genes, suggesting a neural predisposition. Stem Cells Dev. 2006;15:141–64.

Borlongan CV, Kaneko Y, Maki M, Yu SJ, Ali M, Allickson JG, Sanberg CD, Kuzmin-Nichols N, Sanberg PR. Menstrual blood cells display stem cell-like phenotypic markers and exert neuroprotection following transplantation in experimental stroke. Stem Cells Dev. 2010;19:439–52.

Chan RW, Schwab KE, Gargett CE. Clonogenicity of human endometrial epithelial and stromal cells. Biol Reprod. 2004;70: 1738–50.

Hida N, Nishiyama N, Miyoshi S, Kira S, Segawa K, Uyama T, Mori T, Miyado K, Ikegami Y, Cui C, Kiyono T, Kyo S, Shimizu T, Okano T, Sakamoto M, Ogawa

S, Umezawa A. Novel cardiac precursor- like cells from human menstrual blood-derived mesenchymal cells. Stem Cells. 2008;26:1695–704.
Ichim TE, et al. Combination stem cell therapy for heart failure. IntArch Med. 2010;3(1):5.
Ichim TE, et al. Mesenchymal stem cells as anti-inflammatories: implications for treatment of Duchenne muscular dystrophy. Cell Immunol. 2010;260(2):75–82.
Masuda H, Matsuzaki Y, Hiratsu E, Ono M, Nagashima T, Kajitani T, Arase T, Oda H, Uchida H, Asada H, Ito M, Yoshimura Y, Maruyama T, Okano H. Stem cell-like properties of the endometrial side population: implication in endometrial regeneration. PLoS One. 2010;5:e10387.
Meng X, Ichim TE, Zhong J, et al. Endometrial regenerative cells: a novel stem cell population. J Transl Med. 2007;5:57.
Nikoo S, Ebtekar M, Jeddi-Tehrani M, Shervin A, Bozorgmehr M, Kazemnejad S, Zarnani AH. Effect of menstrual blood-derived stromal stem cells on proliferative capacity of peripheral blood mononuclear cells in allogeneic mixed lymphocyte reaction. J Obstet Gynaecol Res. 2012;38:804–9.
Padykula HA. Regeneration in the primate uterus, the role of stem cells. Ann N Y Acad Sci. 1991;622:47–52.
Patel AN, Park E, Kuzman M, et al. Multipotent menstrual blood stromal stem cells, isolation, characterization and differentiation. Cell Transplant. 2008;17:303–11.
Prianishnikov VA. On the concept of stem cell and a model of functional-morphological structure of the endometrium. Contraception. 1978;18:213–23.
Santamaria X, Massasa EE, Feng Y, Wolff E, Taylor HS. Derivation of insulin producing cells from human endometrial stromal stem cells and use in the treatment of murine diabetes. MolTher. 2011; 19(11):2065–71.
Schüring AN, Schulte N, Kelsch R, Röpke A, Kiesel L, Götte M. Characterization of endometrial mesenchymal stem-like cells obtained by endometrial biopsy during routine diagnostics. Fertil Steril. 2011;95:423–6.
Schwab KE, Gargett CE. Co-expression of two perivascular cell markers isolates mesenchymal stem-like cells from human endometrium. Hum Reprod. 2007;22(11):2903–11.
Wolff EF, Gao XB, Yao KV, Andrews ZB, Du H, Elsworth JD, Taylor HS. Endometrial stem cell transplantation restores dopamine production in a Parkinson's disease model. J Cell Mol Med. 2011;15:747–55.
Zhong Z, et al. Feasibility investigation of allogeneic endometrial regenerative cells. J Transl Med. 2009;7:15.

3.2 Applications in Ischemic Heart Disease

Ischemic heart disease, myocardial infarction, and other vascular diseases are the leading cause of death and constitute a major epidemiological challenge. They are a major threat to human health nowadays. Therapeutic approaches mostly aim to restore blood flow to a localized segment by angioplasty or bypass surgery. However, those treatments are temporary. Vascular regeneration through therapeutic angiogenesis and/or arteriogenesis describes a strategy where blood vessel formation is induced for the purposes of treatment and/or prevention of ischemic disease and holds the promise of effective permanent treatment. Thus, application of novel therapeutic approaches involving stem cells has gained considerable attention in recent years. Stem cells and endothelial progenitor cells can differentiate into vascular lineages that repair vascular systems. The rationale of stem cell therapy of cardiac infarct was to implant cells, which will be able to transdifferentiated into cardiomyocytes and regenerate the cardiac muscle. The obvious candidates for treating these diseases were stem cells obtained from bone marrow. The applications of bone marrow-derived cells for heart failures and its related diseases are enticing. This is due to its directed cardiomyocyte differentiation and its ability to secrete angiogenic and trophic factors.

In 2001, cell-based therapy for cardiac diseases was reported using autologous myoblast and bone marrow mononuclear cells. Bone marrow stem cells were both used to treat ischemic heart disease and myocardial infarction. A study compared bone marrow mononuclear cells with allogeneic bone marrow MSC in 30 patients with ischemic cardiomyopathy. Their follow-up showed no difference between two groups, with a mild increase in ejection fraction. Bone marrow has a proven track record as stand-alone treatment in patients with ischemic heart disease and myocardial infarction. However, angiogenic potency of bone marrow in patients with coronary artery disease is impaired. Besides, use of bone marrow is limited by the need for anesthesia during the bone marrow extraction procedure, which is dangerous in the ischemic heart failure people. Thus, adipose tissue was also demonstrated for its in vivo experimentation. In 2004, the cardiomyogenic potential of ADSC has been documented. The same mechanisms allow using ADSC for treatment of animal model of severe hind limb ischemia. However, obtaining adipose-derived stem cells is not easy, and it is still an invasive procedure, and it could not be a readily available source all the time.

Endometrial stem cells could outweigh the obstacles and many other such hindrances faced by other post-natal stem cells, thereby entering the clinic in an efficient manner. With the dynamic cyclical regenerative and angiogenic potency of endometrial stem cells, it could be concluded that endometrial stem cells could be an ideal source of treating cardiovascular disorders. There are also proven track record on its efficacy in treating vascular disorders. As stated above, endometrium undergoes rapid angiogenesis in a controlled manner every month. With this in-built potency, upregulated production of angiogenic factors including PDGF, EGF, and VEGF and MMPs have been described both in the mouse and human endometrium and its production is stimulated monthly by estradiol.

Studies demonstrated myocytic differentiation of ERC-like cells. These results demonstrate that the endometrial progenitor cells and menstrual blood-derived cells can transfer dystrophin into dystrophied myocytes. Based on this study, a pioneering study on the therapeutic use of MenSCs in an animal model of post-infarct cardiac injury was performed by Hida et al. in 2008. The researchers isolated MenSCs, transfected them with an enhanced green fluorescent protein (EGFP) adenoviral construct (for detection purposes only), and achieved cardiocytic differentiation using a coculture system with embryonic cardiomyocytes. They demonstrate that menstrual blood-derived mesenchymal cells exhibit cardiomyocyte-specific action potentials. They also observed a superior rate of post-infarct recovery of ejection fraction, as well as reduction in fibrosis with the ERC-like cells in an animal model. Overall, the data of this study open up exciting therapeutic perspectives for cardiac stem cell-based therapies.

By modifying the original protocol of Hida et al., Ikegami et al. were able to perform a cardiomyogenic differentiation of MenSCs under serum-free conditions. Murine fetal cardiomyocytes were initially cultured as feeders on laminin-coated plates with a cardiomyocyte differentiation medium. Successful cardiomyogenic differentiation was demonstrated under serum-free condition compared to serum-containing media. This has created an important advancement in therapeutics, as for applications in humans, the development of serum-free differentiation protocols represents an ideal option. However, this study demands further investigations in pre-clinical models.

A great breakthrough was obtained by Medistem group by performing a phase II clinical trial in which cardiac patients are treated with stem cells obtained from menstrual blood. The RECOVER-ERC congestive heart failure Phase II clinical trial is testing the safety of the ERC (endometrial regenerative cells) proprietary stem cell population using a 30-min catheter-based retrograde delivery technique, through which the stem cells are administered. ERC are administered without tissue matching or the requirement for immune suppressive drugs. They had treated many patients with no adverse events. Thus, these stem cells are potent source for treating ischemic heart disease.

Further Reading

Al Sabti H. Therapeutic angiogenesis in cardiovascular disease. J Cardiothorac Surg 2007;2:49.

Cui CH: Menstrual blood-derived cells confer human dystrophin expression in the murine model of Duchenne muscular dystrophy via cell fusion and myogenic transdifferentiation. *Mol Biol Cell* 2007, 18(5):1586–1594.

Hamano K, Nishida M, Hirata K, Mikamo A, Li TS, Harada M, Miura T, Matsuzaki M, Esato K. Local implantation of autologous bonemarrow cells for therapeutic angiogenesis in patients with ischemic heart disease: clinical trial and preliminary results. Jpn Circ J 2001;65:845–7.

Hamano K. Local implantation of autologous bone marrow cells for therapeutic angiogenesis in patients with ischemic heart disease: clinical trial and preliminary results. *Jpn Circ J* 2001, 65(9):845–847.

Hare JM: Comparison of Allogeneic vs Autologous Bone Marrow-Derived Mesenchymal Stem Cells Delivered by Transendocardial Injection in Patients With Ischemic Cardiomyopathy: The POSEIDON Randomized Trial. *JAMA* 2012, 12;308(22):2369–2379.

Hida N, Nishiyama N, Miyoshi S, Kira S, Segawa K, Uyama T, Mori T, Miyado K, Ikegami Y, Cui C, Kiyono T, Kyo S, Shimizu T,Okano T, Sakamoto M, Ogawa S, Umezawa A. Novel cardiac precursor- like cells from human menstrual blood-derived mesenchymalcells. Stem Cells. 2008;26:1695–704.

Ikegami Y, Miyoshi S, Nishiyama N, Hida N, Okamoto K, MiyadoK, Segawa K, Ogawa S, Umezawa A. Serum-independent cardiomyogenictransdifferentiation in human endometrium-derived mesenchymalcells. Artif Organs. 2010;34:280–8.

Menasche P: Myoblast transplantation for heart failure. *Lancet* 2001, 357(9252):27 280.

Meyer GP: Intracoronary bone marrow cell transfer after myocardial infarction: 5-year follow-up from the randomized-controlled BOOST trial. *Eur Heart J* 2009, 30(24):2978–2984

Miniño AM, Murphy SL, Xu J, Kochanek KD. Deaths: final data for 2008. Natl Vital Stat Rep. 2011;59:1–126.

Miyahara Y, Nagaya N, Kataoka M, Yanagawa B, Tanaka K, HaoH, Ishino K, Ishida H, Shimizu T, Kangawa K, Sano S, Okano T, Kitamura S, Mori H. Monolayeredmesenchymal stem cells repair scarred myocardium after myocardial infarction. Nat Med. 2006;12:459–65.

Nakagami H, Maeda K, Morishita R, Iguchi S, Nishikawa T, Takami Y, Kikuchi Y, Saito Y, Tamai K, Ogihara T, KanedaY. Novel autologous cell therapy in ischemic limb disease through growth factor secretion by cultured adipose tissue-derived stromal cells. ArterioscierThrombVasc Biol. 2005;25:2542–7.

Niklaus AL: Effect of estrogen on vascular endothelial growth/permeability factor expression by glandular epithelial and stromal cells in the baboon endometrium. *Biol Reprod* 2003, 68(6):1997–2004.

Planat-Benard V, Silvestre JS, Cousin B, Andre M, Nibbelink M, Tamarat R, Clergue M, Manneville C, Saillan-Barreau C, DuriezM, Tedgui A, Levy B, Penicaud L, Casteilla L. Plasticity of human adipose lineage cells toward endothelial cells: physiological and therapeutic perspectives. Circulation. 2004;109:656–63.

Suzuki K: Targeted cell delivery into infarcted rat hearts by retrograde intracoronary infusion: distribution, dynamics, and influence on cardiac function. *Circulation* 2004, 110(11 Suppl 1):II225-II230.

Tateno K, Minamino T, Toko H, Akazawa H, Shimizu N, Takeda S, Kunieda T, Miyauchi H, Oyama T, Matsuura K, Nishi J, Kobayashi Y, Nagai T, Kuwabara Y, Iwakura Y, Nomura F, Saito Y, Komuro I. Critical roles of muscle-secreted angiogenic factors in therapeutic neovascularization. Circ Res 2006;98:1194–202.

van der Bogt KEA, Schrepfer S, Yu J, Sheikh AY, Hoyt G, GovaertJA, Velotta JB, Contag CH, Robbins RC, Wu JC. Comparison of transplantation of adipose tissue- and bone marrow-derived mesenchymal stem cells in the infracted heart. Transplantation.2009;87:642–52.

Walter DH: Impaired CXCR4 signaling contributes to the reduced neovascularization capacity of endothelial progenitor cells from patients with coronary artery disease. *Circ Res* 2005, 97(11):1142–1151.
Wang L, Deng J, Tian W, Xiang B, Yang T, et al. Adipose-derived stem cells are an effective cell candidate for treatment of heart failure: an MR imaging study of rat hearts. Am J Physiol Heart Circ Physiol. 2009;297:1020–31.
Winter A, Breit S, Parsch D, Benz K, Steck E, Hauner H, Weber RM, Ewerbeck V, Richter W. Cartilage-like gene expression in differentiated human stem cell spheroids: a comparison of bone marrow-derived and adipose tissue-derived stromal cells. Arthritis Rheum. 2003;48:418–29.
Yousef M: The BALANCE Study: clinical benefit and long-term outcome after intracoronary autologous bone marrow cell transplantation in patients with acute myocardial infarction. *J Am Coll Cardiol* 2009, 53(24):2262–2269.
Zhang DZ, Gai LY, Liu HW, Jin QH, Huang JH, Zhu XY. Transplantation of autologous adipose-derived stem cells ameliorates cardiac function in rabbits with myocardial infarction. Chin Med J (Engl). 2007;120:300–7.

3.3 Applications in Limb Ischemia

Acute hind limb ischemia (ALI) is a medical condition caused by a sudden lack of blood flow to hind limbs. ALI is due to either an embolism or thrombosis. Thus, ALI can be caused by peripheral vascular disease, air, trauma, fat, amniotic fluid, or a tumor. Critical Limb Ischemia (CLI) is an advanced form of peripheral artery disease. This buildup of plaque, also known as atherosclerosis, narrows or blocks blood flow, reducing circulation of blood. It comprises medical conditions such as chronic ischemic rest pain, ulcers, or gangrene caused by proven occlusive disease. Treatment of CLI includes the following: medications; endovascular treatments such as angioplasty, stents, artherectomy, surgical treatments; and finally amputation. Amputation-free survival (AFS) has been recommended as the gold standard for evaluating No-Option Critical Limb Ischemia (NO-CLI) therapy. With the recent technical advancements in regenerative medicine, stem cells and endothelial progenitor cells derived from bone marrow have been investigated as possible treatments for CLI.

This has proven clinically effect and had reduced need for amputation in CLI patients receiving bone marrow or peripheral stem cells, as administration of bone marrow cells in hindlimb ischemia led to neoangiogenesis. In 2002, Tateishi-Yuyama et al. used the technique of bone marrow extraction and mononuclear cell preparation for CLI treatment in 45 patients with a better follow-up with development of proangiogenic cytokines. Other groups also reported similar therapeutic results. Clinical studies using autologous stem cell therapy for CLI has been reviewed in Sprengers et al.'s article. However, this is limited by several disadvantages including the need for anesthesia during the bone marrow extraction procedure, which is dangerous in the CLI population since numerous co-morbidities exist. Besides, an age-associated decline in angiogenic potential of bone marrow cells occurs, and it was observed that patients with peripheral artery disease have a significantly inhibited angiogenic potential as compared with age-matched controls. Besides, ASCs are also gaining consensus since 2001 as it is readily available from autologous adipose tissue, and have significant potential for tissue repair under conditions of myocardial infarction, heart failure, hind limb ischemia, and inflammation. Thus, adipose-derived stem cells were also demonstrated to have effect similar to bone marrow in treating CLI. The same mechanisms allow using ADSC for treatment of animal model of severe hind limb ischemia. However, adipose-derived stem cell also does not support to be an ideal source of transplantation offering no side-effects, obstacles, or non-invasiveness.

Dr Taylor's group demonstrated that various endometrial cells are actually derived from bone marrow. Possibility of using endometrial cells for treating critical limb ischemia as an alternative source has been demonstrated due to its high levels of growth factors and MMPs and highly potent angiogenic activity. Besides, a study showed endometrial cells produce various factors such as VEGF, BDNF, and neurotrophin-3. Moreover, enhanced antifibrotic activity was also confirmed by researchers. Murphy et al. created an aggressive murine limb ischemia model comprising of femoral artery ligation with nerve excision. In a pilot study on 16 mice, 1×10^6 MenSC were injected intramuscularly into the mice. ERC administration was capable of reducing limb loss in all treated animals, whereas control animals suffered limb necrosis. They had demonstrated several safety measures of extending this therapy to patients in their study, and they have registered for a clinical trial evaluating intramuscular administration of ERC in patients with CLI ineligible for surgical or endovascular intervention. The registered interventional phase I/II clinical trial with a ClinicalTrials.gov identifier: NCT01558908. Besides, Ichim group has also registered for clinical trials, and presently, they are conducting phase I clinical trial in CLI with number: NCT01558908 and a phase II double-blind placebo-controlled cardiac study. They have reported safety and efficacy of their cell administration.

Based on the results of Murphy et al., Ngoc Bich Vu et al. have also reported the efficacy of MenScs for treating acute hind limb ischemia. Using the protocol of Niyama et al., Ngoc Bich Vu et al. successfully established grade IV hind limb ischemia in all mice. Later, mice were treated with MenSC transplantation immediately. MenSC transplantation clearly improved ischemia in mouse models. They hypothesized that the MenSCs can take part in these processes by two mechanisms. The first mechanism is that MenSCs stimulate de novo angiogenesis by secreted factors. This is because MenScs were shown previously to produce various factors such as VEGF, BDNF, and neurotrophin-3. The other mechanism relates to the differentiation of MenSCs into endothelial cells that participate in de novo vasculogenesis as MenSCs easily differentiate into endothelial cells when induced in an endothelial induction medium. They also proposed that intramuscular injection results in better improvement than intravenous injection. Moreover, they concluded that combination of intramuscular injection and infusion into a vein may provide the best outcomes with justifications. Thus, with all these aforesaid advantages, it could be predicted that endometrium serves as a better source as compared to other post-natal source for treatment of limb ischemia.

Further Reading

Asahara T, Masuda H, Takahashi T, Kalka C, Pastore C, Silver M, Kearne M, Magner M, Isner JM: Bone marrow origin of endothelial progenitor cells responsible for postnatal vasculogenesis in physiological and pathological neovascularization. *Circ Res*, 1999;85:221–228.

Attanasio S, Snell J: Therapeutic angiogenesis in the management of critical limb ischemia: current concepts and review. Cardiol Rev, 2005;17:115–20.

Borlongan CV, Kaneko Y, Maki M, Yu SJ, Ali M, Allickson JG, Sanberg CD, Kuzmin-Nichols N, Sanberg PR. Menstrual blood cells display stem cell-like phenotypic markers and exert neuroprotection following transplantation in experimental stroke. Stem Cells Dev, 2009:19(4):439–52

Du H, Taylor HS: Contribution of Bone Marrow-Derived Stem Cells to Endometrium and Endometriosis. Stem Cells, 2007:25(8):2082–2086.

Fowkes FG, Rutherford RB, TASC II Working Group: Inter-society consensus for the management of peripheral arterial disease. Int Angiol, 2009:26:81–157.

Girling JE, Rogers PA: Recent advances in endometrial angiogenesis research. *Angiogenesis*, 2005:8:89–99.

Hida N, Nishiyama N, Miyoshi S, Kira S, Segawa K, Uyama T, Moi T, Miyako K, Ikegami Y, Cui C, Kiyono T, Kyo S, Shimuzu T, Okano T, Sakamoto M, Ogawa S, Umezawa A. Novel cardiac precursor- like cells from human menstrual blood-derived mesenchymal. Stem Cells. 2008;26:1695–704.

Ichim TE, et al. Combination stem cell therapy for heart failure. IntArch Med. 2010;3(1):5.

Ichim TE, et al. Mesenchymal stem cells as anti-inflammatories: implications for treatment of Duchenne muscular dystrophy. Cell Immunol. 2010;260(2):75–82.

Jaffery Z, Thornton SN, White CJ. Acute limb ischemia. Am J Med Sci. 2011;342(3):226–34.

Murphy MP, Lawson JH, Rapp BM, Dalsing MC, Klein J, Wilson MG, Hutchins GD, March KL. Autologous bone marrow mononuclear cell therapy is safe and promotes amputation-free survival in patients with critical limb ischemia. J Vasc Surg. 2011;53:1565–74.

Murphy MP: Allogeneic endometrial regenerative cells: an "Off the shelf solution" for critical limb ischemia? *J Transl Med* 2008, **6**:45

Nakagami H, Maeda K, Morishita R, Iguchi S, Nishikawa T, Takami Y, Kikuchi Y, Saito Y, Tamai K, Ogihara T, Kaneda Y. Novel autologous cell therapy in ischemic limb disease through

growth factor secretion by cultured adipose tissue-derived stromal cells. Arterioscler Thromb Vasc Biol. 2005;25:2542–7.

Ngoc Bich Vu, Van Ngoc-Le Trinh, Lan Thi Phi, Ngoc Kim Phan, and Phuc Van Pham (2015). Human Menstrual Blood-Derived Stem Cell Transplantation for Acute Hind Limb Ischemia Treatment in Mouse Models: In Regenerative Medicine (Eds): Niranjan Battacharya and Phillip Stubblefield, 205–218.

Niiyama H, Huang NF, Rollins MD, Cooke JP. Murine model of hindlimb ischemia. J Vis Exp. 2009;(23):e1035.

Nizankowski R, Petriczek T, Skotnicki A, Szczeklik A: The treatment of advanced chronic lower limb ischaemia with marrow stem cell autotransplantation. *Kardiol Pol* 2005, **63**:351–360.

Patel AN, et al. Multipotent menstrual blood stromal stem cells: isolation, characterization, and differentiation. Cell Transplant. 2008;17(3):303–11.

Rebelatto CK, AM Aguiar, MP Moretao, AC Senegaglia, P Hansen, F Barchiki, J Oliveira, J Martins, C Kuligovski, F Mansur, A Christofis, VF Amaral, PS Brofman, S Goldenberg, LS Nakao, A Correa. (2008). Dissimilar Differentiation of Mesenchymal Stem Cells from Bone Marrow, Umbilical Cord Blood, and Adipose Tissue. *Experimental Biology and Medicine* 233(7), 901–913.

Rehman J, Traktuev D, Li J, Merfeld-Clauss S, Temm-Grove CJ, Bovenkerk JE, Pell CL, Johnstone BH, Considine RV, March KL. Secretion of angiogenic and antiapoptotic factors by human adipose stromal cells. Circulation. 2004;109:1292–8.

Setacci C, de Donato G, Teraa M, Moll FL, Ricco JB, Becker F, Robert-Ebadi H, Cao P, Eckstein HH, De Rango P, Diehm N, Schmidli J, Dick F, Davies AH, Lepäntalo M, Apelqvist J. Chapter IV: treatment of critical limb ischaemia. Eur J Vasc Endovasc Surg. 2011;42 Suppl 2:S43–59.

Sprengers RW, Lips DJ, Moll FL, Verhaar MC. Progenitor cell therapy in patients with critical limb ischemia without surgical options. Ann Surg. 2008;247:411–20.

Sprengers RW, Lips DJ, Moll FL, Verhaar MC: Progenitor cell therapy in patients with critical limb ischemia without surgical options. *Ann Surg* 2008, 247:411–420.

Tateishi-Yuyama E, Matsubara H, Murohara T, Ikeda U, Shintani S, Masaki H, Amano K, Kishimoto Y, Yoshimoto K, Akashi H, Shimada K, Iwasaka T, Imaizumi T, Therapeutic Angiogenesis using Cell Transplantation (TACT) Study Investigators: Therapeutic angiogenesis for patients with limb ischaemia by autologous transplantation of bone marrow cells: a pilot study and a randomised controlled trial. *Lancet* 2002, **360**:427–435.

Walker TG. Acute limb ischemia. Tech VascIntervRadiol.2009;12(2):117–29.

Zhong Z, et al. Feasibility investigation of allogeneic endometrial regenerative cells. J Transl Med. 2009;7:15.

3.4 Applications in Diabetes

Over the last few decades, the main therapeutic approach to insulin-dependent diabetes has confined to the use of insulin injections. However, they are not effective in replacing normoglycemic level. Then emerged the transplantation of islets of Langerhans from beta cells of the pancreas. It has its own disadvantage due to shortage of donors and immune rejection. Regeneration of pancreatic β-cell has emerged as a recent advancement in stem cell technology. Insulin-producing beta cell differentiation from stem cells could represent as an attractive alternative. These differentiated beta cells have been transplanted to various animal models of diabetes and had been successful.

Conventionally, bone marrow and subcutaneous adipose tissue are both endowed with the capacity for multi-lineage differentiation, including the possibility of becoming a pancreatic beta cell-like phenotype. However, bone marrow-derived stem cells have a lower proliferative potency and have not proven promising candidates for the treatment of widespread diseases. Despite this limitation, there have been reports on the ability of bone marrow-derived mesenchymal stem cells (BMSC) to successfully transdifferentiate into pancreatic beta cells, thereby indicating the potential of BMSC as a promising alternative in the treatment of diabetes. Adipose-derived stem cells have gained more attention in recent years, for the treatment of type I and type II diabetes. Transdifferentiation of subcutaneous adipose tissue-derived stem cells into beta cells has been carried out in vitro using multistep differentiation procedure. Our group had also demonstrated the potential of bone marrow and adipose tissue-derived stem cells in producing islet-like clusters. We also compared the pancreatic lineage transdifferentiation competence of adipose tissue to that for bone marrow-derived stem cells. Adipose-derived stem cells have been used as a novel therapy for patients suffering type II diabetes, where autologous activated adipose-derived stem cells were given as intravenous or intraperitoneal injections proving its efficacy and usefulness in diabetes treatment (NCT01453751). There have also been studies confirming the efficacy of intravenous administration of autologous activated stromal vascular fraction as a treatment of type II diabetes (NCT00703612 and NCT00703599).

Following this, there are also recent successful studies on endometrial regenerative cells or endometrial stem cells for the treatment of diabetes, due to all its potential benefits mentioned throughout this book. Studies have shown that human endometrial-derived mesenchymal stem cells (hEMSCs) have the ability to treat human diseases including type 1 diabetes. In brief, the use of endometrial stromal-derived insulin-producing cells in mouse model of diabetes has been studied successfully. Besides, phase I/II clinical trial has also been introduced using menstrual blood-derived stem cells with registered clinical trial number NCT01496339. This emphasizes the significance of endometrial stem cells as an ideal tool to treat diabetes. The details of how these stem cells had been used to treat diabetes and its significance are as follows.

Hugh Taylor's group had shown that endometrial mesenchymal stem cells could successfully be reprogrammed into insulin-producing cells in vitro. The study was confirmed by various molecular and histochemical analyses. They used the small molecule indolactam V, to increase the yield of cells that expressed pancreatic β-cells pan-markers. They found these cells as effective in an STZ-induced model of diabetes. Mice from the group that were transplanted with differentiated cells displayed no gross pathological symptoms and showed a stabilization of glucose; in contrast, diabetic mice transplanted with undifferentiated cells showed a large number of complications.

Another study conducted by HY Li and his coworkers also reported the induction of insulin-producing cells from endometrial mesenchymal stem cells. They isolated endometrial MSCs and succeeded in differentiating them into spheroid body endometrial MSCs (SB-EMSCs) under serum-free induction media along with essential growth factors for 2 weeks. They found SB-EMSCs could efficiently differentiate into both insulin- and glucagon-positive cells, expressing high levels of those molecules. cDNA microarray analysis reported that expression profiles of SB-EMSCs are related to those of islet tissues. Furthermore, upon differentiation, SB-EMSCs displayed increased mRNA expression levels of NKx2.2, Glut2, insulin, glucagon, and somatostatin. In vivo xenotransplantation into STZ-induced SCID mice confirmed that SB-EMSC-derived beta cells helped to restore blood insulin levels in diabetic SCID mice and greatly prolonged the survival of graft cells.

Menstrual blood-derived stem/progenitor cells (MBPCs) were also used to treat diabetes in an animal model by another research group. They demonstrated two hypotheses in their study. One is to investigate the therapeutic effect of menstrual blood cells in treating type I diabetes in a mouse model, and second is to study the repair mechanism involved. They demonstrated that intravenous injection of MBPCs ameliorated diabetic symptoms by reversing hyperglycemia, maintaining body weight, recovering islet structure, and stimulating endogenous beta cell regeneration and an increased insulin production. They further analyzed the in vivo distribution of these cells and demonstrated that majority of MBPCs migrated into damaged pancreas and located at the islet, duct, and exocrine tissue and promote endogenous pancreatic progenitor differentiation. They found that menstrual blood-derived cell did not directly differentiate into insulin-producing cells, but enhanced neurogenin3 (ngn3) expression, which represented endocrine progenitors that were activated. They also found a series of genes associated with the embryonic mode of beta cell development. They concluded that the menstrual blood-derived cells stimulated beta cell regeneration through promoting differentiation of endogenous progenitor cells.

These aforesaid results emphasize the significance of menstrual blood and endometrial stem cells in treating diabetes, thereby proving the potential applications of endometrium-derived stem cells in treating diseases.

Further Reading

Abdi R, Fiorina P, Adra CN, Atkinson M, Sayegh MH. Immunomodulation by mesenchymal stem cells: a potential therapeutic strategy for type 1 diabetes. Diabetes. 2008;57: 1759–67.

Berney T, Mamin A, Shapiro J, et al. Detection of insulin mRNA in the peripheral blood after human islet transplantation predicts deterioration of metabolic control. Am J Transplant. 2006;6: 1704–11.

Dhanasekaran M, Indumathi S, Harikrishnan R, Rashmi M, Rajkumar JS et al., (2013). Human Omentum fat derived Mesenchymal stem cells transdifferentiates into pancreatic islet like cluster Cell Biochem Func. 31(7), 612–619.

Dhanasekaran M, Indumathi S, Sudarsanam D and Rajkumar JS. (2013) Differentiation of mesenchymal stem cells derived from human bone marrow and subcutaneous adipose tissue into pancreatic islet-like clusters *in vitro. Cell and Mol Biol Lett* 18(1), 75–88.

Garg S, Rosenstock J, Silverman B, et al. Efficacy and safety of preprandial human insulin inhalation powder versus injectableinsulin in patients with type 1 diabetes. Diabetologia. 2006;49: 891–9.

Li HY, Chen YJ, Chen SJ, Kao CL, Tseng LM, Lo WL, Chang CM, Yang DM, Ku HH, Twu NF, Liao CY, Chiou SH, Chang YL. Induction of insulin-producing cells derived from endometrial mesenchymal stem-like cells. J Pharmacol Exp Ther. 2011;335:817–29.

Oh, S.H., Muzzonigro, T.M., Bae, S.H., LaPlante, J.M., Hatch, H.M. and Petersen, B.E. Adult bone marrow-derived cells trans-differentiating into insulin-producing cells for the treatment of type I diabetes. Lab. Invest. 84 (2004) 607–617.

Ryan EA, Shandro T, Green K, et al. Assessment of the severity of hypoglycemia and glycemic ability in type 1 diabetic subjectsundergoing islet transplantation. Diabetes. 2004;53:955–62.

Santamaria X, Massasa EE, Feng Y, Wolff E, Taylor HS. Derivation of insulin producing cells from human endometrial stromal stem cells and use in the treatment of murine diabetes. Mol Ther. 2011;19:2065–71.

Shapiro AMJ, Lakey JRT, Ryan EA, et al. Islet transplantation in seven patients with type 1 diabetes mellitus using a glucocorticoidfreeimmunosuppressive regimen. N Engl J Med. 2000;343: 230–8.

Shapiro AMJ, Ricordi C, Hering BJ, et al. International trial of the edmonton protocol for islet transplantation. N Engl J Med. 2006;355:1318–30.

Sun Y, Chen L, Hou X.G., Hou W.K., Dong, J.J., Sun, L., Tang, K.X., Wang, B., Song, J., Li, H. and Wang, K.X. Differentiation of bone marrowderived mesenchymal stem cells from diabetic patients into insulinproducing cells in vitro. Chin. Med. J. 120 (2007) 771–776.

Timper K, Seboek D, Eberhardt M, et al. Human adipose tissuederivedmesenchymal stem cells differentiate into insulin, somatostatin, and glucagon expressing cells. BiochemBiophys Res Commun. 2006;341:1135–40.

Toso C, Vallee JP, Morel P, et al. Clinical magnetic resonance imaging of pancreatic islet grafts after iron nanoparticle labeling. Am J Transplant. 2008;8:701–6.

Uccelli A, Zappia E, Benvenuto F, Frassoni F, Mancardi G. Stem cells in infl ammatory demyelinating disorders: a dual role for immunosuppression and neuroprotection. Expert Opin Biol Ther. 2006;6:17–22.

Urbán VS, Kiss J, Kovács J, Gócza E, Vas V, Monostori E, Uher F. Mesenchymal stem cells cooperate with bone marrow cells in therapy of diabetes. Stem Cells 2008, 26:244–253.

Voltarelli JC, Couri CE, Stracieri AB, Oliveira MC, Moraes DA, Pieroni F, Coutinho M, Malmegrim KC, Foss-Freitas MC, Simões BP, Foss MC, Squiers E, Burt RK. Autologous nonmyeloablative hematopoietic stem cell transplantation in newly diagnosed type 1 diabetes mellitus. JAMA 2007; 297:1568–1576.

Wu Xiaoxing, Luo Yueqiu, Chen Jinyang, Pan Ruolang, Xiang Bingyu, Du Xiaochun, Xiang Lixin, Shao Jianzhong, and Xiang Charlie. Stem Cells and Development. 2014, 23(11): 1245–1257.

Zhu, Y., Liu, T., Song, K., Fan, X., Ma, X. and Cui, Z. Adipose-derived stem cell: a better stem cell than BMSC. Cell Biochem. Funct. 26 (2008) 664–675.

3.5 Applications in Stroke

Stroke is the leading cause of mortality and significant morbidity in India (prevalence of approximately 0.5 %) and worldwide. It is ranked as the sixth leading cause of disability and projected to rank fourth by 2020. Despite successful efforts to decrease incidence and mortality of cerebrovascular diseases, they still remain as a major concern in clinical setting. Currently tissue plasminogen activator (TPA) therapy is the only known FDA-approved treatment for stroke. TPA is a thrombolytic agent for the dissipation of clots. Unfortunately, the drug is time-sensitive and ineffective if not administered on time. In fact, in 2008 only 1.8–2.1 % of all patients affected by ischemic strokes in the United States received the therapy. Further studies have tried to evaluate the possibility of extending the limit beyond 3 h, but the results were non-conclusive. Studies demonstrated an increase in mortality with an increased risk of intra-cranial hemorrhagic complications.

Considering these facts, stem cell therapy has become an encouraging approach in the ischemic areas. Although evidence of the beneficial effects of stem cells in animal stroke models is growing, there is lack of enough clinical data. MSC-based therapeutic approaches have been tested in a range of animal models of human diseases, for treatment of conditions such as myocardial infarction, brain and spinal cord traumatic injury, stroke, and fibrosis. Specifically, MSC-based therapy has also been investigated for the treatment of several models of central nervous system diseases, affecting both the brain [traumatic brain injury and cerebral infarct (ischemic stroke)] and the spinal cord (traumatic spinal cord injury). Intravenous administration of bone marrow-derived MSCs into a rat model of stroke was shown to improve functional recovery, increase FGF-2 expression, reduce apoptosis, and promote endogenous cellular proliferation. Several other studies have successfully reported the role of bone marrow-derived MSC therapy for stroke. Furthermore, effects of cord blood stem cells and adipose-derived stem cells were also been enticing in a stroke model. The stimulation of neurogenesis in post-stroke injury models has been demonstrated to be linked directly to angiogenesis. As an example, Borlongan et al. observed, in rat models of stroke, that umbilical cord blood cells were effective on rat models of stroke in promoting repair mechanism, probably through the production of growth factors, cytokines, and other therapeutic molecules.

Borlongo and their team continue to study the effect of menstrual blood stem cells in stroke model as menstrual blood cell injections are proposed as a restorative therapy after stroke. Using a neural induction media (DMEM/F12 containing N2 and FGF-2) and a subsequent stimulation with retinoic acids, the authors could induce expression of the neural markers MAP2 and nestin in the MenSCs. Moreover, under hypoxic conditions, menstrual blood cells provided neuroprotection as in vivo, which leads to significant improvement in motor asymmetry, motor coordination, and neurologic tests. However, except for migration, they did not show signs of differentiation, expressing their original markers. This reveals cell differentiation is not the main pathway of neuroprotection or neuroregeneration.

Rodrigues et al. also used autologous menstrual blood cells in treating subacute phase of stroke. Although women have readily available stem cells throughout the life, the effect of these menstrual blood and endometrial stem cells in treating stroke is still uncertain. Thus, it warrants more experimental data to fully understand the pathophysiological role of MenSC and eMSCs in experimental stroke therapy. However, its success is not too far, and these cells will represent important therapeutic tool that may improve disease outcome, thereby decreasing the mortality rate of stroke patients.

Further Reading

Bliss T, Guzman R, Daadi M, Steinberg GK. Cell transplantation therapy for stroke. Stroke 2007;38 (2 Suppl):817–26.

Boltze J, Kowalski I, Geiger K, et al. Experimental treatment of stroke in spontaneously hypertensive rats by CD34+ and CD34-cord blood cells. Ger Med Sci. 2005;3: Doc09.

Borlongan CV, Hadman M, Sanberg CD, Sanberg PR. Central nervous system entry of peripherally injected umbilical cord blood cells is not required for neuroprotection in stroke. Stroke. 2004;35: 2385–9.

Borlongan CV, Kaneko Y, Maki M, Yu SJ, Ali M, Allickson JG, Sanberg CD, Kuzmin-Nichols N, Sanberg PR. Menstrual blood cells display stem cell-like phenotypic markers and exert neuroprotection following transplantation in experimental stroke. Stem Cells Dev. 2010;19:439–52.

Carpenter CR, Keim SM, Milne WK, Meurer WJ, Barsan WG, TheBest Evidence in Emergency Medicine Investigator Group. Thrombolytic therapy for acute ischemic stroke beyond 3 hours. J Emerg Med. 2011;40:82–92.

Centers for Disease Control and Prevention (CDC). Prevalence of disabilities and associated health conditions among adults, United States. MMWR Morb Mortal Wkly Rep. 1999;50:120–5.

Chen J, Li Y, Wang L, Zhang Z, Lu D, Lu M, et al. Therapeutic benefit of intravenous administration of bone marrow stromal cells after cerebral ischemia in rats. Stroke 2001;32(4):1005–11.

Chen J, Sanberg PR, Li Y, Wang L, Lu M, Willing AE, Sanchez-Ramos J, Chopp M. Intravenous administration of human umbilical cord blood reduces behavioral deficits after stroke in rats. Stroke. 2001;32:2682–8.

Cronin CA. Intravenous tissue plasminogen activator for stroke, a review of the ECASS III results in relation to prior clinical trials. J Emerg Med. 2010;38:99–105.

Dharmasaroja P. Bone marrow-derived mesenchymal stem cells for the treatment of ischemic stroke. J Clin Neurosci 2009;16(1):12–20.

Hacke W, Kaste M, Bluhmki E, et al. Thrombolysis with alteplase3 to 4.5 hours after acute ischemic stroke. N Engl J Med. 2008;359: 1317–29.

Honmou O, et al. Mesenchymal stem cells: therapeutic outlook for stroke. Trends Mol Med. 2012;18(5):292–7.

Kleindorfer D, Lindsell CJ, Brass L, Koroshetz W, Broderick JP. National US estimates of recombinant tissue plasminogen activator use, ICD-9 codes substantially underestimate. Stroke. 2008;39:924–8.

Leu S, Lin YC, Yuen CM, Yen CH, Kao YH, Sun CK, Yip HK. Adipose-derived mesenchymal stem cells markedly attenuate brain infarct size and improve neurological function in rats. J Transl Med. 2010;8:63.

Park DH, Borlongan CV, Willing AE, et al. Human umbilical cord blood cell grafts for brain ischemia. Cell Transplant. 2009;18: 985–98.

Parr AM, Tator CH, Keating A. Bone marrow-derived mesenchymalstromal cells for the repair of central nervous system injury. Bone Marrow Transplant. 2007;40:609–19.

Rodrigues MC, Glover LE, Weinbren N, Rizzi JA, Ishikawa H, Shinozuka K, Tajiri N, Kaneko Y, Sanberg PR, Allickson JG, Kuzmin-Nichols N, Garbuzova-Davis S, Voltarelli JC, Cruz E, Borlongan CV. Toward personalized cell therapies: autologous menstrual blood cells for stroke. J Biomed Biotechnol. 2011;2011:194720.

Wu J, Sun Z, Sun HS, Wu J, Weisel RD, Keating A, Li ZH, Feng ZP, Li RK. Intravenously administered bone marrow cells migrate to damaged brain tissue and improve neural function in ischemic rats. Cell Transplant. 2008;16:993–1005.

Zawadzka M, Lukasiuk K, Machaj EK, Pojda Z, Kaminska B. Lack of migration and neurological benefi ts after infusion of umbilical cord blood cells in ischemic brain injury. Acta Neurobiol Exp. 2009;69:46–51.

3.6 Applications in Parkinson's Disease

In 1817, James Parkinson documented six cases he had been observing in "An Essay on The Shaking Palsy," describing the classic motor symptoms of the disease. Thus, the disease is named after him. Parkinson's disease is a chronic progressive degenerative disease of the central nervous system. It is caused by a breakdown of dopamine in the substantia nigra. Dopamine is a neurotransmitter that stimulates the motor neurons that in turn control muscles. When dopamine production is reduced, the nerves are not able to control movement or maintain coordination. Although researchers are trying to understand this disease in detail, what triggers degeneration of the dopaminergic neurons remains unknown. Current drug therapy or surgeries have their limitations as they focus on symptomatic relief only and do nothing to slow down the disease. While drug therapy and surgery do not fundamentally solve the problem, transplanting dopaminergic neurons into the brains of patients with Parkinson's disease yields more promising results. Studies on fetal neural transplantations and neural stem cell transplantation were successful. The study demonstrated that transplanted dopaminergic neurons could resolve the issue. However, there are limitations with transplantation of fetal tissue including ethical issues.

Differentiating stem cells into dopamine-producing neurons and transplanting these cells into patients provide symptomatic relief. Neural stem cell transplantation therapy is also successful in regenerating dopaminergic neurons. NSCs influence the existing CNS, rescuing the present neurons and recovering the striatal dopaminergic system. However, it is a very invasive procedure. Thus, emerged the isolation of adult stem cells from other sources such as bone marrow, adipose, cord blood and so on. Induced pluripotent stem cells are gaining more importance in treating Parkinson's diseases and showed improved motor symptoms in Parkinson's disease model rats. However, the success it is still under consideration.

Stem cells derived from the endometrium and transplanted into the brains of laboratory mice with Parkinson's disease appear to restore functioning of brain cells. The findings are published in the Journal of Cellular and Molecular Medicine. Although these are preliminary results, the findings increase the likelihood that endometrial tissue could be harvested from women with Parkinson's disease and used to re-grow brain areas that have been damaged by the disease, according to lead author Hugh S. Taylor, M.D., Professor in the Department of Obstetrics, Gynecology & Reproductive Sciences at Yale School of Medicine, and Section Chief of Reproductive Endocrinology and Infertility at Yale School of Medicine. Taylor and his colleagues collected and cultured endometrial tissue from nine women and verified that they could be transformed into dopamine-producing nerve cells like those in the brain. They exhibited axon projections, pyramidal cell bodies, and dendritic projections that recapitulate synapse formation; these cells also expressed the neural marker nestin and tyrosine hydroxylase. A whole cell patch clamp recording method was used to characterize the cells. These cells not only survived in the location they are transplanted, but also spontaneously migrated to damaged areas and spontaneously differentiated in vivo. Taylor also points out that endometrial stem cells are one of the best sources for generating neurons because they appear to less likely to be rejected than stem cells from other sources. Engraftment was demonstrated up to 5 weeks following transplantation.

Following this study, recently the team has also investigated their potential for clinical cellular therapies using a non-human primate model in a pilot feasibility study. They demonstrate that endometrium-derived stem cells may be transplanted into an MPTP-exposed monkey model of PD. After injection into the striatum, endometrium-derived stem cells engrafted, exhibited neuron-like morphology, expressed tyrosine hydroxylase (TH), and increased the numbers of TH-positive cells on the transplanted side in vivo. However, further extensive research is mandatory to make this successful in human Parkinson's diseases.

Further Reading

Alexi T, Borlongan CV, Faull RL, et al. Neuroprotective strategies for basal ganglia degeneration: Parkinson's and Huntington's diseases. Prog Neurobiol. 2000;60:409–70.

Arenas E. Towards stem cell replacement therapies for Parkinson's disease. Biochem Biophys Res Commun. 2010;396:152–6.

Björklund A, Cenci MA. Recent advances in Parkinson's disease: basic research. Prog Brain Res. 2010;183:ix–x

Glavaski-Joksimovic A, Virag T, Chang QA, et al. Reversal of dopaminergic degeneration in a parkinsonian rat following micrografting of human bone marrow-derived neural progenitors. Cell Transplant. 2009;18:801–14.

Lindvall O, Rehnerona S, Brundin P, et al. Human fetal dopamine neurons grafted into the striatum in two patients with severe Parkinson's disease. a detailed account of methodology and a 6-month follow-up. Arch Neurol. 1989;46:615–31.

Madrazo I, Francobourland R, Ostrosky-Solis F, et al. Fetal homotransplant to the striatum of parkinsonian subjects. Arch Neurol. 1990;47:1281–5.

Parkinson's disease. a detailed account of methodology and a 6-month follow-up. Arch Neurol. 1989;46:615–31.

Venkataramana NK, Kumar SK, Balaraju S, et al. Open-labeled study of unilateral autologous bone-marrow-derived mesenchymal stem cell transplantation in Parkinson's disease. Transl Res. 2010; 155:62–70.

Wernig M, Zhao J, Pruszak J, et al. Neurons derived from reprogrammed fibroblasts functionally integrate into the fetal brain and improve symptoms of rats with Parkinson's disease. Proc Natl Acad Sci U S A. 2008;105:5856–61.

Wolff EF, Gao XB, Yao KV, et al. Endometrial stem cell transplantation restores dopamine production in a Parkinson's disease model. J Cell Mol Med. 2011;15:747–55.

Yasuhara T, Matsukawa N, Hara K, Yu G, Xu L, Maki M, Kim SU, Borlongan CV. Transplantation of human neural stem cells exerts neuroprotection in a rat model of Parkinson's disease. J Neurosci. 2006;26:12497–511.

Chapter 4
Threats and Challenges of Endometrial Stem Cells

Endometrial stem cells are already known for its higher proliferation, differentiation, ability to undergo fast angiogenesis during menstruation, and immune tolerance for embryo during pregnancy. Thus, they are considered as a valuable source of stem cells. Several researches have demonstrated the multi-differentiation potency of endometrial MSCs both in vitro and in vivo. They had demonstrated to differentiate into various cell types such as insulin producing cells, osteoblasts, neurons, myoblast, and chondrocytes, which are detailed in earlier chapters. Besides, pre-clinical and clinical experimental trial and therapies had also become successful in treating various diseases such as myocardial infarction, stroke, Parkinson's disease, and diabetes as discussed earlier. However, on the other side of the coin, it possesses a threat to us by means of causing several gynecological disorders due to its abnormal proliferation and differentiation. It is postulated that several gynecological conditions are associated with abnormal endometrial proliferation, and it is possible that putative endometrial stem/progenitor cells may play a role in the pathophysiology of diseases such as endometriosis, endometrial hyperplasia, endometrial cancer, and adenomyosis (Fig. 4.1). Alterations in the number, function, regulation, and location of epithelial/stromal endometrial stem/progenitor cells may be responsible for any one of these endometrial diseases. Furthermore, study of the clinical correlations of endometrial stem cells with gynecological diseases may unravel several unresolved barriers and lead to the use of endometrial stem cells as an ideal alternative source of curative therapeutics. Thus, addressing this barrier would be a great challenge and beneficial for the womankind. The following section will address the role of endometrial stem cells in causing endometrial disorders such as endometriosis, adenomyosis, endometrial hyperplasia, and endometrial carcinoma. Therefore, endometrial stem cells could be considered as "dual role payer" or a "double-edged sword"; wherein, on one side, it plays a dynamic role in natural function of endometrium, on the other side, it contributes to the pathogenesis of various endometrial disorders through its abnormal proliferation and differentiation.

Currently, two major problems hinder the medical assistance to the women suffering from endometrial disorders: the lack of appropriate biomarkers useful in

I. Somasundaram, *Endometrial Stem Cells and Its Potential Applications*, SpringerBriefs in Stem Cells, DOI 10.1007/978-81-322-2746-5_4

Endometrial Disorders	
Endometrial hyperplasia/Cancer	Mutated stem/progenitor – tumor responsible for progression, metastasis, recurrence
Endometriosis	Normal stem/progenitor cell shed into peritoneal cavity – ectopic implant
Adenomyosis	Normal stem/progenitor cells, abnormal niche, inappropriate differentiation–ectopic growth, SMC hyperplasia
Asherman's Syndrome	Damage/loss of normal stem/ progenitor cells
Inadequate endometrium for IVF	Diminished activity of normal stem/progenitor cells
	Gargett ,2007 13: 87-101

Fig. 4.1 Human endometrium and its disorders

early diagnosis and the inexistence of conservative (medical) treatments with long-lasting effect. Further characterization of the stem cell populations is challenging and may open up possibilities for rational development of novel and improved diagnostic and therapeutic strategies.

4.1 Endometriosis

Endometriosis is a chronic benign gynecological disease which is characterized by the ectopic formation of endometrial stroma and glands mostly seen in pelvic peritoneum. It is found in about 11 % of all women population during reproduction. Comparative microarray analysis of gene expression in patients with ectopic and eutopic endometrial cells has demonstrated difference in gene expression pattern between the two. Despite its common occurrence, the pathogenesis of endometriosis is poorly understood. A number of theories have been proposed in its pathogenesis: genetic and environmental factors, immune system, retrograde menstruation, coelomic metaplasia, embryonic rest theory, lymphovascular metastasis, and stem cell-based theory. The most widely accepted mechanism is Sampson's retrograde menstruation theory.

However, recently, stem cell-based theory is gaining much attention in causing endometriosis, and it has been considered to be major players in the pathogenesis of endometriosis and other endometrium-associated diseases. This is due to the fact that endometrial stem/progenitor cells are inappropriately shed during menstruation and reach the peritoneal cavity where they adhere and establish endometriotic implants. Numerous studies demonstrate that unfractionated human endometrial cells establish ectopic endometrial growth in the many experimental models. Some study demonstrated recruitment of bone marrow-derived mesenchymal stem cells to the endometrium to point out the role of stem cells at the pathogenesis of endometriosis.

The evidential proof of the above is as follows: some stemness-related genes are preferentially expressed in endometriosis lesions (e.g., *UTF1*, *TCL1A*, *ZFP42*, and *SALL4*) and other undifferentiation markers (e.g., *GDF3*). Mechanism determining the self-renewal rate and stem cell fate are dysregulated in endometriosis. Consequently, altered stem cell behavior, chromosomal aberrations, impaired DNA methylation, histone modifications, and imbalance of microRNA (miRNA) expression are associated with the phenotypic changes of endometriosis. Overall, there are five main functional alterations involved: increased proliferation, migration, adhesion/invasiveness, pro-angiogenic factor production, and dysregulated expression of certain immune modulators. The involvement of stem cells in pathogenesis of endometriosis and their dysfunction have become increasingly evident. The strategies that will certainly deserve an in-depth consideration of targeting molecular mechanism of stem cells in endometriosis-regulating recruitment, adhesion, survival, self-renewal, and pro-angiogenic behavior.

Further Reading

Buck Louis, G. M., M. L. Hediger, et al. (2011). "Incidence of endometriosis by study population and diagnostic method: the ENDO study." Fertility and sterility. 96(2): 360–365

Taylor, R. N., S. G. Lundeen, et al. (2002). "Emerging role of genomics in endometriosis research" Fertility and sterility 78(4): 694–698.

Giudice, L. C. (2003). Genomics' role in understanding the pathogenesis of endometriosis, 21(2): 119–124.

Giudice, L. C. (2006). "Application of functional genomics to primate endometrium: insights into biological processes." Reprod Biol Endocrinol 4(suppl 1): S4.

Gazvani R, Templeton A. New considerations for the pathogenesis of endometriosis. Int J Gynaecol Obstet. 2002;76:117–26.

D' Hooghe TM, Debrock S, Meuleman C, et al. Future directions in endometriosis. Obstet Gynecol Clin North Am. 2003;30:221–44.

Figueira PG, Abrão MS, Krikun G, Taylor HS, Taylor H. Stem cells in endometrium and their role in the pathogenesis of endometriosis. *AnnNY AcadSci* (2011) **1221**:10–7.doi:10.1111/j.1749-6632.2011.05969.x

Deane JA, Gualano RC, Gargett CE. Regenerating endometrium from stem/progenitor cells: is it abnormal in endometriosis, Asherman;s syndrome and infertility? *CurrOpinObstetGynecol* (2013); 25(3):193–200.

Starzinski-Powintz A, Zeitvogel A, Schreiner A, et al. In search of pathogenic mechanism in endometriosis: the challenge for molecular cell biology. Curr Mol Med. 2001;1:655–64.

Sasson IE, Taylor HS. Stem cells and the pathogenesis of endometriosis. Ann My Acad Sci. 2008;1127:106–15.
Esfandiari, N., J. Ai, et al. (2007). "Expression of Glycodelin and Cyclooxygenase 2 in Human Endometrial Tissue Following Three dimensional Culture." American Journal of Reproductive Immunology 57(1): 49–54.
Esfandiari, N., M. Khazaei, et al. (2007). "Effect of a statin on an in vitro model of endometriosis." Fertility and sterility 87(2): 257–262.
Ferenczy, A. and C. Bergeron (1991). "Histology of the human endometrium: from birth to senescence." Annals of the New York Academy of Sciences 622(1): 6–27.
Figueira, P. G. M., M. S. Abrão, et al. "Stem cells in endometrium and their role in the pathogenesis of endometriosis." Annals of the New York Academy of Sciences 1221(1): 10–17.
Fujishita, A., A. Hasuo, et al. (2000). "Immunohistochemical study of angiogenic factors in endometrium and endometriosis." Gynecologic and obstetric investigation 48(1): 36–44.
Taylor, H. S. (2004). "Endometrial cells derived from donor stem cells in bone marrow transplant recipients." JAMA: the journal of the American Medical Association 292(1): 81–85.
Forte A, Schettino MT, Finicelli M, Cipollaro M, Colacurci N, CobellisL, etal. Expression pattern of stemness-related genes in human endometrial and endometriotict issues. *MolMed* (2009) **15**(11–12):392–401.
Giudice LC, Kao LC. Endometriosis. *Lancet* (2004) **364**(9447):1789–99. doi:10.1016/S0140-6736(04)17403–5.
Gilabert-Estelles J, Braza-Boils A, Ramon LA, Zorio E, Medina P, Espana F, etal. Role of microRNAs in gynecological pathology. *Curr Med Chem* (2012) **19**(15):2406–13.doi:10.2174/092986712800269362
Kao AP, Wang KH, Chang CC, Lee JN, Long CY, Chen HS, et al. Comparative study of human eutopic and ectopic endometrial mesenchymal stemcells and the development of an in vivo endometriotic invasion model. *Fertil Steril* (2011) **95**(4):1308 15.e1.

4.2 Endometrial Hyperplasia

Endometrial hyperplasia is an abnormal proliferation of the endometrium. It is a risk factor for the development of endometrial carcinoma. The presence of unopposed estrogen – which, for example, may result from exogenous estrogen therapy, anovulatory cycles, polycystic ovary syndrome (PCOS), or obesity – has been shown to increase the likelihood of developing endometrial hyperplasia and cancer. Patients with endometrial hyperplasia are always accompanied with abnormal bleeding; this may be menorrhagia, menometrorrhagia, or post-menopausal bleeding. Endometrial cancer must be ruled out particularly in patients older than 35

Type	Description	Risk of Progression to Endometrial Cancer
Simple	Dilated glands that may contain some outpouching and abundant endometrial stroma	~ 1%
Complex	Glands are crowded with very little endometrial stroma, and a very complex gland pattern and outpouching formations	~ 3%-5%
Simple with atypia	Is the same as above, but also contains cytologic atypia. This refers to hyperchromatic, enlarged epithelial cells with an increased nuclear to cytoplasmic ratio.	~ 8%-10%
Complex with atypia		~ 25%-30%

Fig. 4.2 WHO classification of categories of hyperplasia

years of age who present with these conditions. The most common method used to diagnose hyperplasia and cancer is endometrial aspiration, the endometrial biopsy.

The important aspect of treatment of hyperplasia is the classification of the type of hyperplasia. Endometrial hyperplasia is defined as a proliferation of glands of irregular size and shape with an increase in the glands/stroma ratio. WHO classifies four categories of risk classification: simple hyperplasia, complex hyperplasia, simple hyperplasia with atypia, and complex hyperplasia with atypia (Fig. 4.2). We can subdivide them into mild atypia (nuclear enlargement and rounding with evenly dispersed chromatin) or moderate atypia (clumped chromatin, larger nuclear size, prominent nucleoli). Hyperplasia without atypia, either simple or complex, rarely (1 %, 3 %) progresses to carcinoma. In contrast, atypical endometrial hyperplasia is believed to be the direct precursor to endometrioid carcinoma. Endometrial hyperplasia with atypia is most likely to progress to type 1 endometrial carcinoma, which accounts for 97 % of uterine cancers, whereas simple hyperplasia rarely progresses to carcinoma. Excess of estrogen relative to progesterone is the known risk factors for endometrial hyperplasia. Therefore, progestin is used to treat endometrial hyperplasia. However, in the presence of atypia, the ideal management is hysterectomy.

Several kinds of genetic alterations had been detected in endometrial cancers, including PTEN inactivation, beta-catenin (CTNNB1) mutations, microsatellite instability, and activational mutations of the K-ras gene. The concept of EIN and the diagnostic schema was introduced by Mutter and the Endometrial Collaborative Group in 2000 and later launched at Brigham and Women's Hospital in 2002, to replace the older hyperplasia-based nomenclature by WHO, which implies endometrial hyperplasia as the precancerous lesion of type I cancers. According to this model, the premalignant lesions are referred to as EIN to distinguish them from the diffuse estrogen associated changes of benign endometrial hyperplasia.

Thus, to summarize, there are currently two systems of endometrial pre-cancer nomenclature in common usage: the 1994 4-Class WHO schema (WHO94) and the EIN diagnostic schema. The WHO94 is based upon a seminal, albeit small, and retrospective study in 1985 by Kurman and colleagues, which correlated cytological atypia with increased risk for cancer. These categories are descriptive in nature, and interpretation is subjective; accordingly, studies indicate poor reproducibility of the individual case classification. In the schema developed by International Endometrial Collaborative Group, endometrial pre-cancers are termed endometrial intra-epithelial neoplasia; histomorphological, genetic, clinical, and biological data were used to develop quantitative pathologic criteria for three disease categories of benign, premalignant, and malignant disease. Several molecular alterations and microRNAs were identified in endometrium under normal and disease progression condition. One of the previous works in this area focused on the normal hormonal regulation of human endometrium toward understanding molecular deregulation occurring in pathological situations and also may expose novel therapeutic opportunities. One of the study established concomitant differential miRNA and mRNA expression profiles of uterine epithelial cells purified from endometrial biopsy specimens in the late proliferative and mid-secretory phases. The study identified 12 miRNAs (MIR29B, MIR29C, MIR30B, MIR30D, MIR31, MIR193A-3P, MIR203, MIR204, MIR200C, MIR210, MIR582-5P, and MIR345) whose expression was significantly up-regulated in the mid-secretory-phase samples were predicted to target many cell cycle genes. Thus, they suggest a role for miRNAs in down-regulating the expression of some cell cycle genes in the secretory-phase endometrial epithelium, thereby suppressing cell proliferation. Many mRNAs have been shown to be differentially expressed in the rat uterus during embryo implantation, including work on pre-receptive and receptive stage. Scientists have focused on miRs that are differentially regulated during window period and receptivity affecting cell cycle and apoptosis. Of great interest in targeting cancers and knowing the regulation of miRs affecting cancer genes, there are many studies on miRNA regulation of cancer and cancer stem cells, including its regulation of epithelial-mesenchymal transition. Although the abnormal proliferation of endometrial stem cells in endometrial cancers is identified, there is hardly any study to correlate the role of endometrial stem cells in endometrial hyperplasia cases.

Further Reading

Artacho-Pérula, E.M. *et al.* Histomorphometry of normal and abnormal endometrial samples. *Int. J. Gynecol. Pathol.*, *12*: 173–179, 1993.

Ausems, E.W.M.A. *et al.* Nuclear morphometry in the determination of the prognosis of marked atypical endometrial hyperplasia. *Int. J. Gynecol. Pathol.*, *4*: 180–185, 1985.

Baak, J.P.A. Further evaluation of the practical applicability of nuclear morphometry for the prediction of the outcome of atypical endometrial hyperplasia. *Anal. Quant. Cytol. Histol.*, *8*: 46–48, 1986.

Baak, J.P.A. The role of computerized morphometric and cytometric feature analysis in endometrial hyperplasia and cancer prognosis. *J. Cell. Bioch.*, Supplement 23:137–146, 1995.

Kurman, R. J., P. F. Kaminski, et al. (1985). "The behavior of endometrial hyperplasia. A long-term study of "untreated" hyperplasia in 170 patients." Cancer 56(2): 403–12.

Mutter, G. L. (2000). "Endometrial intraepithelial neoplasia (EIN): will it bring order to chaos? The Endometrial Collaborative Group." Gynecol Oncol 76(3): 287–90.
Mutter, G. L. (2002). "Diagnosis of premalignant endometrial disease." J Clin Pathol 55(5): 326–31.
Mutter, G. L., M. C. Lin, et al. (2000). "Altered PTEN expression as a diagnostic marker for the earliest endometrial precancers." J Natl Cancer Inst 92(11): 924–30.
Mutter, G. L., R. J. Zaino, et al. (2007). "Benign endometrial hyperplasia sequence and endometrial intraepithelial neoplasia." Int J Gynecol Pathol 26(2): 103–14.
Scully RE, B. T., et al., Ed. (1994). Histological Typing of Female Genital Tract Tumors Uterine corpus. New York, NY, Springer Verlag.
Tashiro, H., M. S. Blazes, et al. (1997). "Mutations in PTEN are frequent in endometrial carcinoma but rare in other common gynecological malignancies." Cancer Res 57(18): 3935–40.

4.3 Endometrial Carcinoma

Endometrial cancer (EC) is the most common gynecological malignancy. Type I endometrial cancers have a relatively favorable prognostic profile compared to type II endometrial cancer. Mainly cases are diagnosed at an early stage due to the readily detectable symptom of abnormal uterine bleeding in post- and peri-menopausal women. In the last two decades, there have been significant advances made in the study of the precursors of type I endometrial cancer, and this precancerous lesion is currently considered as endometrial atypical hyperplasia in the WHO classification system (that is still the most frequently used by pathologists) and the "endometrial intra-epithelial neoplasia" (EIN) system that was originally proposed by Mutter et al. as explained under Sect. 4.2. On the other hand, studies of type II endometrial cancer precursors have been relatively limited.

Several kinds of genetic alterations had been detected in type 1 endometrial cancers, including PTEN inactivation, beta-catenin (CTNNB1) mutations, and to a lesser degree, microsatellite instability (related to inactivation of the MLH1 gene), and activational mutations of the K-ras gene. Several other genes and miRs have been identified. Of great interest in targeting cancers and knowing the regulation of miRs affecting cancer genes, there are many studies on miRNA regulation of cancer and cancer stem cells, including its regulation of epithelial-mesenchymal transition. A great breakthrough in regulation of miRs affecting endometrial adenocarcinoma was identified. Studies showed that several miRNAs are differentially expressed in endometrial adenocarcinoma as compared to normal endometrium. These identified miRNA hold great potential as target for classification and prognosis of this disease. Further analysis of the differentially expressed miRNA and their target genes will help to derive new biomarkers that can be used for classification and prognosis of endometrial adenocarcinoma.

Besides, cancer stem cells play a key role in endometrial carcinoma given that a rare population of epithelial stem/progenitor cells has been identified in human endometrium. It is possible that these cells or their progeny may be the source of the putative CSCs. The first evidence for CSCs was suggested in 1997 in a study on a uterine carcinosarcoma-derived cell line. The highly proliferative clonal cells expressed immunohistochemical and molecular markers consistent with their parental tissue, recapitulated the tumor phenotype in vitro, and were considered stem cells responsible for propagating the cell line. While these results were promising, they were not performed on freshly isolated EC cells, or even on a pure EC cell line. Thus, further studies are warranted and functional assays of CSC activity are required to confirm its role in endometrial carcinoma.

Further Reading

Bethesda, MD: National Cancer Institute. http://seer.cancer.gov/csr/1975_2006/

Chan R, Schwab K & Gargett C 2004 Clonogenicity of human endometrial epithelial and stromal cells. Biology of Reproduction 70 1738–1750.

Cohn D, Fabbri M, Valeri N, Alder H, Ivanov I, Liu C, Croce C, Resnick K: Comprehensive miRNA profiling of surgically staged endometrial cancer. Am J Obstet Gynecol 2010, 202:656.e-.e8.

Creasman WT, Odicino F, Maisonneuve P, Quinn MA, Beller U, Benedet JL, Heintz APM, Ngan HYS & Pecorelli S 2006 Carcinoma of the corpus uteri. International Journal of Gynecology and Obstetrics 95 S105–S143.

Esteller, M., L. Catasus, et al. (1999). "hMLH1 promoter hypermethylation is an early event in human endometrial tumorigenesis." Am J Pathol 155(5): 1767–72.

Friel AM, Sergent PA, Patnaude C, Szotek PP, Oliva E, Scadden DT, Seiden MV, Foster R & Rueda BR 2008 Functional analyses of the cancer stem cell-like properties of human endometrial tumor initiating cells. Cell Cycle 7 242–249.

Gargett CE 2007 Uterine stem cells: what is the evidence? Human Reproduction Update 13 87–101.

Gargett CE, Chan RWS & Schwab KE 2008 Hormone and growth factor signaling in endometrial renewal: role of stem/progenitor cells. Molecular and Cellular Endocrinology 288 22–29.

Gorai I, Yanagibashi T, Taki A, Udagawa K, Miyagi E, Nakazawa T, Hirahara F, Nagashima Y & Minaguchi H 1997 Uterine carcinosarcoma is derived from a single stem cell: an in vitro study. International Journal of Cancer 72 821–827.

Hao, J. et al. (2014) MicroRNA control of epithelial-mesenchymal transition in cancer stem cells. Int. J. Cancer 135, 1019–1027.

Horner MJ, Ries LAG, Krapcho M, Neyman N, Aminou R, Howlader N, Altekruse SF, Feuer EJ, Huang L, Mariotto A et al. (Eds) 2009 SEER Cancer Statistics Review, 1975–2006.

Tashiro, H., M. S. Blazes, et al. (1997). "Mutations in PTEN are frequent in endometrial carcinoma but rare in other common gynecological malignancies." Cancer Res 57(18): 3935–40.

Velasco, A., E. Bussaglia, et al. (2006). "PIK3CA gene mutations in endometrial carcinoma: correlation with PTEN and K-RAS alterations." Hum Pathol 37(11): 1465–72.

Wu W, Lin Z, Zhuang Z, Liang X: Expression profile of mammalian microRNAs in endometrioid adenocarcinoma. Eur J Cancer Prev 2009, 18:50–55.

Xia HF, Jin XH, Cao ZF, Hu Y, Ma X, MicroRNA expression and regulation in the uterus during embryo implantation in rat. FEBS J (2014). 281(7); 1872–91.

Chapter 5
Conclusion and Future Prospects

The uterus is the largest female reproductive organ that plays an integral role in implantation and in absence of pregnancy, menstruation. Human endometrium, lining the uterine cavity, is a simple columnar epithelium that exhibits dramatic cyclical changes during each menstrual cycle. Studies on therapeutic applications of these endometrial stem cells in treating various disorders at both pre-clinical and clinical settings are gaining consensus in recent years. This is due to its high vasculature, extensive proliferation coupled with inherent role of angiogenesis in the reproductive process such as embryo implantation and endometrial regeneration after menstruation. However, endometrial stem cells are a "double-edged sword." It has both therapeutic implication and establishes itself likely to cause diseases due to dysregulatory mechanism. Thus, first three chapters of the brief summarized the angelic property, that is, the potential in vitro and in vivo applications of endometrial stem cells, and the last chapter summarizes the demonic property of endometrial stem cells, which is an important aspect of uterine stem cell biology. This is because recent studies have proven that abnormal proliferation and differentiation of endometrial stem cells are responsible for the dysregulated endometrial mechanism. Especially, the major concerns of endometrial disorders include endometriosis, endometrial hyperplasia, and endometrial carcinoma.

This brief provides the overlay of the human endometrium and its derived stem cells of all aspects at a single roof. In detail, this Springer Brief presents an understanding of stem cells derived from human endometrium and examines how cells from these sources are being used in vitro and in vivo research therapies and treatments for various clinical problems and diseases. This acts as a base for further development of endometrial stem cell research that might be of worthwhile investigation in the near future to expand this benchside research to bedside not only in certain diseases mentioned in this book, but also other varying diseases. Further investigating these attributes in pathologic conditions of endometrium might develop more effective therapy for infertility, endometriosis, endometrial hyperplasia, and carcinoma through a targeted therapeutic approach.

I. Somasundaram, *Endometrial Stem Cells and Its Potential Applications*, SpringerBriefs in Stem Cells, DOI 10.1007/978-81-322-2746-5_5

In continuation with my past research work on identifying the in vitro attributes of endometrial stem cells under long-term culturing conditions, our research at D.Y. Patil University presently focuses on the identification of molecular regulatory profiles of human endometrium at normal and disease states, especially under hyperplasia conditions. This paves way to achieve differential regulated common gene profiles of hyperplasia and carcinoma confining to a better therapeutic approach. Besides, other research work to identify the miRNA regime at both tissue and cellular level under normal and disease conditions is also underway. This study will allow us to determine cellular functions and molecular pathways targeted by these differentially expressed miRNAs.

To conclude, undoubtedly, endometrial stem cells may become key players in regenerative medicine because of their pre-natal embryologic origin, redundancy, ease of isolation, dynamic proliferative, and multitude differentiation property. Thus, endometrial stem cells, with its dynamic potency, could be a connecting link between embryonic and other post-natal stem cells. By means of its stemness, endometrial stem cells serves a better tool in achieving cell-based therapies. Besides, further research on endometrial diseases specified above with regards to endometrial stem cells will ease the endometrial disorders to a great extent.

Printed in the United States
By Bookmasters